P9-CQK-225

THE BEAUTY OF DISCOMFORT

The Beauty of Discomfort

How What We Avoid Is What We Need

AMANDA LANG

Collins

The Beauty of Discomfort

Published by Collins, an imprint of HarperCollins Publishers Ltd

First edition

HarperCollins books may be purchased for educational, business, or sales promotional use through our Special Markets Department.

HarperCollins Publishers Ltd
2 Bloor Street East, 20th Floor
Toronto, Ontario, Canada
M4W 1A8

www.harpercollins.ca

Library and Archives Canada Cataloguing in Publication information is available upon request

ISBN 978-1-44344-984-7

Printed and bound in the United States
LSC/H 10 9 8 7 6 5 4 3 2

For Geoff

CONTENTS

INTRODUCTION

Imagine an industry that is decades old, stable and reasonably profitable, and serving customers in a reliable way. Service has been standardized so that customers get roughly the same experience no matter where they are in the country or which provider they choose. It's a highly regulated industry, so companies offering the service need a bunch of licences to operate and must comply with strict rules about employees' qualifications and the number of hours they're permitted to work per day. In many places, there are even regulations about where each company can offer its services so that all providers get a "fair" slice of the pie. The stability of the industry means that operating licences have become quite valuable, and owners can sell them for a tidy profit when they want to get out of the business. For this reason, the balance of power has shifted toward business owners and away from customers (who really have no meaningful choices, because all companies are pretty much the same) and workers (who don't share in the industry's profit in any meaningful way). But hey, that's the way it's been for ages, so nobody really questions it.

Until, out of the blue, an upstart company comes along, wanting to offer the same kind of service but in a whole new way. The upstart doesn't bother getting the same licences, doesn't worry about the traditional regulations governing who can be hired, and refuses to steer clear of any particular geographical area. In other words, there's a brashness to the rookie: it doesn't play nice with the incumbents. But customers immediately embrace it, not least because it's cheaper and more convenient. And it's also tipping the balance back toward workers, so their share of the profit goes up.

That's the story of ride-hailing service Uber.

Canadian-born Garrett Camp came up with the idea for the company when he was standing on a snowy street corner in Paris with his friend Travis Kalanick, trying to hail a taxi. It was an impossible task, but rather than just grouch about it to each other as most of us would, the two men—who happened to be successful entrepreneurs on the hunt for a new business idea—asked themselves whether there was a better way to match available cars to riders. They had a hunch that the current system was rife with inefficiencies, and that technology might be the answer.

Camp had already founded one start-up, an Internet search engine called StumbleUpon, and he was still running it (but had sold the business to eBay for $75 million) when he stumbled upon the idea for Uber. He and Kalanick created Uber Cab in early 2009 with $200,000 of their own money; after a trial period in San Francisco, the company, now named Uber, officially launched in 2011. By May, the company was expanding into a new American city every month; within a year, international cities were being added.

Uber doesn't own cars or employ drivers—it's an app that connects riders and drivers over the Internet, automatically billing riders' credit cards. A five-star rating system lets both riders and drivers grade each other; those with persistently poor ratings are drummed out. And the real genius of Uber is that ordinary citizens can use their own cars and fill the gaps when demand jumps.

When the bubble that had insulated the taxi industry for decades popped, existing operators were furious. Pre-Uber, municipal regulations had limited the number of cabs on the street, so taxi companies were always assured a certain amount of business. Uber had changed all that, and taxi revenues were plummeting. So was the value of an operating licence, which in 2012 had cost C$360,000 in Toronto and north of US$1 million in New York. Owners who'd shelled out that much cash were livid; the value of their investment was depreciating daily. Now there were thousands of new vehicles on the street, driven by people who hadn't had to pay for taxi licences.

Cab drivers in cities around the world staged massive protests, but Uber continued to grow. Leaked documents suggest it racked up $3.6 billion in fares (before drivers were paid) in the first half of 2015. When Saudi Arabia's sovereign wealth fund invested another $3.5 billion in the company the following year, Uber's valuation—at least on paper—reached $25 billion.

Overly protected and overly regulated, the taxi industry never really had a chance, right? Only it did. Any one of the incumbents could have created a smartphone app to let customers summon and pay for a taxi or limousine in exactly the same easy, seamless way Uber does. They certainly scrambled to do so afterward, when it was too late. And they could also have managed the peak-demand problem better, putting more cars on the road so that even in a snowstorm, riders would have a better shot at getting a cab.

The problem wasn't that improvements couldn't be implemented but that the taxi industry had become much too comfortable. Complacent, actually. Adaptability just wasn't in its DNA. There was no game plan for how to respond to new information, new customers, or new threats. And the story of an overly regulated, fat, and lazy industry being disrupted by an upstart could just as easily be told about the telecom and banking industries in many countries. The pattern described by Harvard professor Clay Christensen in *The Innovator's Dilemma* is remarkably similar in diverse businesses: new entrants get into a sector at the low end, a part of the market that the giant incumbents don't really care about, and begin offering inferior but cheaper products, sometimes featuring a new technology. (Think of payment systems like Square or the first Android mobile phones.) But as the upstarts innovate and improve their products, they migrate up-market, until they are stealing big, valuable customers from the incumbents, who start to take notice. By that point, however, it's often—in fact, almost always—too late. The incumbent is already on the way out.

Not long ago, I asked Christensen why businesses [illegible]tinue to fall into this trap, despite knowing that disruption [illegible] inevitable. He explained that no CEO can justify threatening or disrupting his main business on the chance that a new one will emerge. It goes against everything that CEO learned in business school, and also goes against human nature.

To understand change, then, we can't just look at industries or individual companies. The human element is everything. Often, one person can make a big difference, as Garrett Camp did. An individual taxi driver could also have lobbied to implement some of the relatively simple but user-friendly changes that led to Uber, years before Camp was standing on that street corner in Paris, wondering when a cab would turn up. But no one did.

Why not? Why do some people drive change while others are blindsided by it? And why are some people able to adapt and thrive when change is forced upon them, while others are flattened by it?

The ability to adapt is more important than ever because technology and globalization have accelerated the pace of change so dramatically. An innovation like the driverless car—unimaginable to our great-grandparents, who probably remembered sharing the road with horses—will be ho-hum to our kids.

Economies that are still focused on yesterday are already paying a price for it. And inevitably, so are individuals. Any North American who's banking on the return of high-paying manufacturing jobs—or a secure job at a newspaper, or a long-term gig in the oil industry—is already struggling, or likely will be in the not-so-distant future. As our world evolves, so must we. Those who can change—and especially, those who can innovate—have the best chance of succeeding.

Several years ago, I became so convinced that reigniting our natural curiosity would unleash innovation, I wrote a book about it. But it's become clear to me that asking *why* questions is only the beginning. It's also the fun part. The best, truest answer to any searching question—whether it's "Why aren't people coming to our stores anymore?" or "Why have I gained ten pounds?"—isn't usually one you want to hear because it's almost always going to challenge you to do something differently. Acting on answers—executing and actually changing—is hardest of all, even if the potential payoff is huge. Anyone who's tried to embark on a new exercise regimen or launch a new product line or rethink a business strategy knows just how difficult change is. Many of us try but fail, and conclude that we can't do it. We know *why* we should change, but we can't figure out *how* to do it. Change is just too hard.

How can we make change easier? As a business journalist I instinctively look at the big picture, so when I set out to search for answers, I started by looking at industry patterns and the stories of individual companies. How did Microsoft face the formidable challenge posed by Apple? How did Intel pivot, even though pivoting involved massive layoffs and a wholesale restructuring? How did Netflix resurrect its fortunes when customers revolted after a fee increase?

I soon realized that I was looking for answers in the

wrong places, because industries and companies don't decide to change or innovate. People do. To understand how to make change easier, I needed to look at individuals, to understand the human dimension that Clay Christensen had identified.

For most of us, change is not comfortable. Sometimes the discomfort comes from uncertainty or ambiguity. *What if we do this and sales tank? What if I don't succeed?* Sometimes it's related to a sense of loss. Change usually demands that you give something up, and whether that something is an HR policy or carbohydrates, the prospect is rarely pleasant. Sometimes the discomfort is even more fundamental, and linked to identity: If I make this change—get divorced, say, or quit my job to start my own business—who will I be? I'll have to reinvent myself.

Many of us conceive of discomfort as a temporary stage, a necessary if unpleasant rite of passage on the way to a better future. Scrimping and saving, living in a crummy apartment, working long hours for too little money—that's all stuff you put up with while waiting for the day when you've "made it." Once you've reached your goal, you can relax a little. Being comfortable *is* the goal.

What's striking about creative and innovative people, whether they're change agents in Silicon Valley or artists whose work is shown at the Venice Biennale, is that comfort doesn't seem to suit them. Some respond to success the way other people respond to failure: they redouble their efforts, working even harder and longer. Many seek out a new challenge altogether; the CEO of a start-up takes up the guitar, the prize-winning novelist starts volunteering at orphanages in Haiti. They seek out new experiences—change—again and again.

Something about that seems to jump-start creativity and innovation. "Evidence shows that creative contributions depend on the breadth, not just depth, of our knowledge and experience," according to a recent article in the *New York Times*. "In fashion, the most original collections come from directors who spend the most time working abroad. In science, winning a Nobel Prize is less about being a single-minded genius and more about being interested in many things. Relative to typical scientists, Nobel Prize winners are 22 times more likely to perform as actors, dancers or magicians; 12 times more likely to write poetry, plays or novels; seven times more likely to dabble in arts and crafts; and twice as likely to play an instrument or compose music."

Now, these scientists may be lousy dancers and piano players, but that's not the point. The point is that a lot of highly successful creative types continually seek out the very kinds of challenges that make the rest of us, including me, pretty uncomfortable. Whether it's straining to complete a triathlon or write a poem, they are willing to go far outside their comfort zone, and that seems to enhance their creativity and also make them more successful in their day jobs. It's not just about having hobbies. It's about seeking out new experiences that force them to stretch.

Here's the thing: truly successful people don't merely tolerate discomfort—they embrace it, seeking it out again and again. And their comfort with discomfort is what makes them so good at change. They seem to experience discomfort as a positive rather than a negative force, and they find a way to use it to motivate themselves to achieve.

To understand how to change, then, we have to understand how to withstand discomfort. This is a very personal project because discomfort, always, is internal. It's mental and emotional and sometimes physical, and leaning into it requires us to use muscles we may not even know we have. Flexing them—allowing for ambiguity, for the unknown, for possibilities that feel risky and maybe downright terrifying—is how creativity and fresh starts become possible.

In the following pages you will meet people who have very different ways of coping with discomfort, but all of them believe that withstanding it has helped them change in ways that have been hugely positive. They include business founders and university students, top athletes and couch potatoes, meditation gurus and military leaders. Some were forced into discomfort through no choice of their own—a life-altering illness, a business fiasco—while others signed up for it because they had goals they were determined to achieve.

What they taught me is this: if you want to change something big, like a company, you have to start small, by changing yourself and becoming more comfortable with discomfort. Fortunately, there are ways to train ourselves to do this, as well as a lot of social science explaining why we should.

Now, I'm aware that discomfort is a hard sell. No one needs to be convinced that there's beauty in comfort—we all know that a five-star hotel room is pretty great. Discomfort doesn't hold out the same promise of pleasure or reward.

If there's gratification, it's going to be delayed. And in the meantime, you need to believe—as people who are virtuosos at change do—that even if you're never rewarded, some degree of discomfort is inherently good for you. It can spur you on, pushing you to test your own limits; it can make you feel as if you're growing and learning; it can give you a whole new perspective on yourself, and on life.

Learning to tolerate, and then embrace, discomfort is the foundation for change, for individuals and businesses alike. To some, it seems to come pretty naturally; others have no choice. Those of us who want to change in some way will need to develop a new set of instincts to respond to discomfort. The good news is that we can do that. But the best news of all is that doing so won't just make us more resilient and more successful, however we define success. It will also make us happier.

CHAPTER ONE

The Benefits of Discomfort

Human beings are wired to seek comfort.

Babies instinctively react to discomfort by signalling distress. When hungry, they cry. Wet? They wail. Cold? Howl. Physical discomfort is usually the primary trigger, but it isn't long before loneliness, boredom, frustration, jealousy, and other feelings unrelated to survival are signalled through tears as well.

A generation or two ago, parents were told to let their children cry, to toughen them up. This clearly went against the grain, or parents would not have needed stern reminders to resist the urge to hold and comfort their babies. In fact, parents seem to be wired to respond to an infant's distress by attempting to figure out what's wrong and remedy it (if only to stop the noise). And these days, experts encourage us to heed those instincts, lest the baby grow up feeling insecure. It's only when a parent decides to try to sleep-train a child that anyone frames discomfort as being for the baby's own good—and even then, many parents' resolve (mine included) crumbles after a minute or two of heart-rending wailing.

What's new, though, is that the desire to protect our kids from discomfort of any sort sometimes persists long after they're able to sleep through the night. The term "helicopter parenting" dates back to 1969, and by the 2000s it was such a commonly used phrase that it had earned a place in the dictionary. Today, hovering over toddlers at the park is the least of it. Some parents go on to write essays for their university-age children and step in to solve even the most minor problems on campus, according to university administrators.

The impulse driving this kind of behaviour is as old as time: we don't want our kids to suffer as we did, even if "suffering" merely involved pulling the occasional all-nighter to get a term paper written. We want our children to have better lives than we did. But these days that's a really tall order, because it's harder to get into university, harder to find a great job, and harder to buy a first house than ever before. No wonder we hover over our children as if they're rare birds, trying to ensure they thrive in this brave new world.

But what if we have it wrong? What if some degree of discomfort is good for kids?

It seems evident that it is. Isn't that why children have chores and homework? Of course, shouldering responsibility isn't as comfortable as lolling around the house, texting with friends, but most people agree that it's quite a bit more beneficial in terms of equipping kids for real life.

In fact, it's difficult to imagine how anyone could achieve much in life without figuring out how to tolerate some degree of discomfort and push past it. As new research into the consequences of helicopter parenting suggests, trying to protect kids from life's turmoil appears to have the opposite effect: it may weaken their coping skills and ramp up their anxiety. More generally, when kids aren't pushed out of their comfort zone and encouraged to try new things—whether it's zucchini or a new way of holding a baseball bat —they miss the opportunity to learn, not just about the world but about themselves and their own capacity to change.

I began to think about this more when my own son started school. Preventing bullying was, and still is, a major endeavour at schools, and I was grateful for that. I certainly didn't want my kid to be pushed around or made to feel inferior. But the laudable goal of preventing playground terrorism seemed to have morphed into an earnest attempt to prevent *any* behaviour that might make another person feel uncomfortable. Responding in any way to "difference," my son was told, was unacceptable. Inclusion and tolerance were good; exclusion and rejection—even principled rejection—were bad. By implication, kids who were snubbed because they transgressed norms—the grandiose liars and nose-pickers, the ball hogs and snobs—were recast as victims of bullying.

I began to wonder about the unintended consequences of trying to shield children from all forms of negative social interaction. Might that not wind up creating more problems—and more serious ones—than it prevents? In the rush to protect kids from any form of hardship—physical, emotional, or mental—would we inadvertently "protect" them from positive formative experiences too?

My own childhood taught me that discomfort can be instructive, in a good way. Growing up with six brothers and sisters, I learned early about boundaries and what happens if you don't respect them. Not unlike puppies in a litter, we tumbled over each other and meted out some fairly rough justice. The two sisters closest to me in age—Elisabeth and my twin sister, Adrian—taught me pretty early that if I didn't play by the rules of the game, the consequences could be very unpleasant. For some reason that I hope our mother doesn't take too personally, no one wanted to be the mom when we played house (the fun was all in how the "kids" in the game tormented the "adults"). One time too often, I decided the game should end just as my turn to be mother rolled around . . . and as a result, found myself shut out of that game, and all others, for a while. Given that my sisters were my closest friends and playmates, being ostracized definitely hit home and led to lasting behavioural change.

I want my son to be able to tolerate discomfort and learn from it without being scarred by it. The idea isn't to "make a man" out of him but to help him acquire a skill that seems to be essential to creativity and change. If you cannot tolerate discomfort, you cannot get better at things that are difficult. You cannot achieve your own goals. You cannot grow.

Even in organized sports, though, the objective seems to be to make kids as comfortable as possible. My son's first soccer league, for instance, was proud of its feel-good policy: the purpose of the game, the kids were told, was not to compete with others but to have fun (competition, apparently, is not fun). Games would not be scored, because keeping score only made some people feel like losers and others like winners. And there were rules to prevent the most talented players from dominating a match: they were required to pass, whether passing made sense or not, and if they made three goals, they were pulled from the game. The best kids were in effect penalized for their superiority, and their play was restricted in cross-league competitions—which were called "festivals," not "tournaments." (It was amusing to note, however, that while officially there was no score, every six-year-old on the field knew exactly how many goals had been scored and which team walked away the winner.)

Maybe all this sounds harmless and a little silly, but I think the impact of these early experiences is more profound than we imagine. Sure, my kid didn't toss and turn at night, racked with envy of other players or shame about his own performance. But he also didn't push himself harder to practise or to step up his own game. Why bother, when goals didn't really count at the "festival"?

The impact on the kids who actually were standouts was likely even greater. Research shows that children who are gifted, whether physically or intellectually, may be even more afraid to risk failure than those who are "average." That's a problem when failure is so often the early result of trying to learn something new or change in some other way. Researchers have found that gifted children, who experience success early, often exhibit less confidence, possibly because they're aware that they didn't exert much effort to achieve results. What they haven't learned is how to try hard at things, fail, then pick themselves up and try again—essential skills not just in school but in life. In other words, they are less likely to be comfortable with discomfort, which may

help explain why prodigies rarely go on to become creative geniuses. Ellen Winner, a psychologist who has studied child prodigies, found that "only a fraction of gifted children eventually become revolutionary adult creators." They can—and often do—perform very well in their chosen fields, but few are able to make the "painful transition" to adulthood and become grown-ups who create something wholly new. What is painful, exactly, about that transition? The fact that the prodigy, long accustomed to approval and gold stars, must get comfortable with discomfort: uncertainty, wrong turns, the possibility of failure, and the certainty of the occasional disappointment.

Students who periodically occupy the discomfort zone learn the habit of making an effort—and learn also that failure is usually the entry fee for attempting to change and grow. However, educators have become increasingly wary about making students feel uncomfortable—even on university campuses, whose raison d'être is to challenge young people to stretch their minds. At schools and universities across North America, educators are being warned against teaching material that might upset their students. Trigger warnings are supposed to be issued before potentially offensive material is taught. What's on the list of such material? F. Scott Fitzgerald's *The Great Gatsby*, which, apparently, contains misogynistic material that might upset women and trigger traumatic flashbacks. I'm not making this up. In the *Atlantic* magazine in September 2015, constitutional lawyer Greg Lukianoff and social psychologist Jonathan Haidt describe an atmosphere in which college professors are terrified of their own students. So quick to take offence are the students that professors are reluctant to say anything provocative at all. But however difficult self-censorship is for the professors, it is almost certainly worse for their students. "It presumes an extraordinary fragility of the collegiate psyche," the authors write, and transforms campuses from safe places for free speech into "'safe spaces' where young adults are shielded from words and ideas that make some uncomfortable."

In other words, schools are working so hard to avoid *giving* offence that they cannot teach one of the most important lessons of young adulthood: how not to *take* offence.

Learning to cope with discomfort—disagreement, aggression, and criticism (even wholly unprovoked aggression and criticism)—can be one of the greatest accomplishments of a child's early school years. Robert Reich, who was among other things President Bill Clinton's undersecretary of labor, attributes much of his personal development and even his later professional success to being bullied as a child. Born with a congenital issue that limited his growth (he may be an intellectual giant, but Reich is only four foot ten), he attracted an outsized portion of bullying. Rather than be cowed by it, however, Reich set about finding ways to offset it. He developed his intellect, for one thing, and formed alliances with larger boys who were willing to protect him. Those friendships wound up being pivotal, partly because those boys had special qualities of their own, such as kindness and courage. Ultimately, being pushed around on the playground shaped Reich in a deep and lasting way, helping to strengthen his resolve to persevere and overcome. To make something of himself.

If he had it to do over, Reich would probably not choose to be wrapped in cotton wool and protected from all the bumps and bruises of childhood. Nor would he volunteer to be bullied. But if forced to choose between those two extremes, he'd be more likely to select the option that created discomfort. Discomfort, after all, forged his character and propelled him to achieve.

Short of bullying or turning ourselves into drill sergeants, how can we as parents help our kids learn to tolerate discomfort so they can grow? Is there any way to make this palatable to kids?

Yes, says Ilana Ben-Ari, the founder of Toronto-based Twenty One Toys. Long-limbed with a slightly ethereal air, the thirtysomething industrial designer doesn't look hard-hearted, but she specializes in making toys that rub kids' noses in discomfort. Nicely. It's good for them, she says with a smile. "Play is super important in teaching us to be adaptable and okay with the unknown," she explains. "Innovation is uncomfortable. We don't want things to change—it goes against our makeup. So step number one is just to be comfortable with discomfort."

Ben-Ari's best-selling product is the Empathy Toy, which consists of two identical sets of five beautifully crafted wooden pieces with a range of textures, shapes, and configurations. There are cog and arrow shapes—some with grooves and notches, some with flat or textured surfaces, some with raised dots evocative of Braille. The pieces can be joined together in thousands of ways, but they don't just glide together. You have to fiddle with them and jiggle them around. That's by design: the point is for players to struggle a little bit.

The Empathy Toy is actually more of a game than a toy, and it involves three players with distinct roles. One person, the observer, assembles a set of pieces into a configuration, then hands it to a blindfolded second player, the guide. The guide feels the configuration and then, using words only, tries to help the third player, the builder—also blindfolded—construct exactly the same shape using the second set of pieces. The game is over when the builder has replicated the original structure. Or almost over. The observer, who has had to bite her tongue and watch silently as the temporarily blind try to lead the temporarily blind, often starts a conversation about the guide's and the builder's false starts and hilarious misunderstandings. All of this usually takes ten to twenty minutes, then players swap roles, blindfolds are donned once more, and it's time for another round.

Ben-Ari believes that the abilities kids need in order to innovate—empathy, comfort with improvisation, and acceptance of failure as a key part of the creative process—cannot be learned via traditional pedagogical methods. Schools, in her opinion, are virtually useless when it comes to teaching kids how to be more creative and innovative. "School is about schedules. The worst thing you can do in school is hand something in late. Not hand in a bad idea—just late." Her most receptive audience for this message? Schools. The Empathy Toy has been shipped to more than eight hundred schools in thirty-five countries.

I was taken with Ben-Ari's passion and vision, but I wondered if the Empathy Toy wasn't one of those creations that sounds great to adults and is a dud with kids. So I asked if I could road-test it on a ski weekend with my son, my sister, her two kids, and one of their friends. I wondered what our video-obsessed, attention-span-challenged offspring would make of a wooden toy that required patience and a back-and-forth that went beyond monosyllabic grunts. I was particularly interested in knowing what my eleven-year-old son, Julian, who can get frustrated when he's not immediately good at something, would think of it.

To my surprise, the kids couldn't get enough of the game. They vied for the chance to be not only the builder but also the observer, since putting the toy together even without a blindfold on is a creative, fun endeavour. Interestingly, the kids were considerably better at being the observer than the adults were; we couldn't stop jumping in to try to "help" the builder with hints (common, Ben-Ari says, especially for teachers and nurses, who seem to have the hardest time watching children struggle). But Julian and the other kids didn't seem to mind a bit that the game was hard. That's what they *liked* about it. I also think the toy was a hit because success demanded teamwork. The guide really wanted the builder to understand the instructions, and the builder really wanted to follow the instructions and assemble the thing correctly. Amazingly, our screen-obsessed kids didn't mind sitting quietly and watching two players talk each other through a complicated construction.

Watching the kids play with the toy, I was struck again by how misguided it is to try to shelter children from all discomfort, or to present it as a negative rather than a challenge. Not only can they cope with a little discomfort when it's framed as a challenge—especially a playful challenge—but they will build their confidence and stick-to-itiveness as they work through it. Kids who play with Ben-Ari's toy don't just develop empathy—they also foster resilience, collaboration, and creativity.

Those "soft skills," prerequisites for innovation, are more important now than ever before—and they're increasingly difficult to nurture in our instant-gratification world of gaming and texting. Robin Wilson, a guidance counsellor at St. John's High School in Winnipeg, told me some kids refuse to wear the blindfold when playing with the Empathy Toy—it makes them feel too vulnerable. Others get so irritated when they can't assemble the toy with ease that they whip off their blindfolds and hurl the toy's pieces across the room. Always on the lookout for a teachable moment, Wilson uses their reactions to encourage her students "to take a temperature check" and rate their anger or frustration on a five-point scale. "One, you're happy and relaxed; five, you're ready to explode and your brain has stopped working properly." For her, the Empathy Toy is about developing not just empathy but self-regulation, by helping kids identify and manage their emotions so they can recognize when they're escalating to three and can avoid hitting five and having a meltdown.

Ben-Ari's latest toy may come in handy in that regard: the Failure Toy helps kids become comfortable with failing, and to view it as a necessary step in the learning process. Next up is the Improv Toy, so they get comfortable taking risks. Ben-Ari's mission—to equip kids with the twenty-first-century skills they'll need to venture into their discomfort zones and realize they can thrive there—is all about the upside of discomfort.

There's a significant downside if kids don't learn to get comfortable with discomfort, and it goes far beyond a reduced capacity for growth and innovation. People who can't tolerate discomfort can wind up with much more serious problems. Just ask Judson Brewer, an expert on addiction.

In the early nineties, Brewer, now the director of research at the Center for Mindfulness at the University of Massachusetts Medical School, began thinking about the mind–body connection—an idea that at the time was still viewed as New Agey, bordering on crackpot. In particular, Brewer, then an undergraduate at Princeton, was curious about why people get sick when they're stressed. It had happened to him, and he'd noticed that antibiotics didn't provide a cure.

When he entered a joint MD/PhD program at Washington University in St. Louis, suffering from insomnia after a bad breakup, he happened upon Jon Kabat-Zinn's book *Full Catastrophe Living: Using the Wisdom of Your Body and Mind to Face Stress, Pain, and Illness.* It was an idea Brewer could relate to at that particular stressed-out moment, when his life seemed to be falling apart. The book promised that mindfulness—"Paying attention, on purpose, in the present moment, and non-judgmentally"—could help, and included detailed instructions on mindfulness meditation. Brewer was game to try anything that might make the pain stop.

So he began meditating. And meditating, and meditating. The more he did, the more clearly he began to see how he was contributing to his own stress, and what that stress was doing to him. He found that mindful meditation—simply sitting and focusing on his breath, then eventually on his thoughts and emotions (but without any judgment)—helped him step back and observe his own behaviour more objectively. He got over his heartbreak. He found himself better able to refrain from snapping when he was sleep-deprived and more empathetic with his patients.

Eight years later, doctorate and medical degree in hand, Brewer headed to Yale to begin training as a psychiatrist, and there he began researching how mindfulness can affect the brain and help treat psychiatric conditions, particularly addictions. Addiction—whether to Krispy Kreme donuts, crack cocaine, or Twitter—is one of the most "evolutionarily conserved learning processes currently known in science," Brewer explained in his 2015 TED Talk. The human brain is old school, or at least its reward-based learning system is. Even though we're no longer roaming the savannah dodging predators and trying to score our next meal, our brains are still wired as though we are, and that's why seductive, relatively modern substances—alcohol, sugar, and nicotine, to name a few—can literally "hijack" the brain's dopamine-based reward system.

We see a cupcake that looks good and our brain says, "Calories are good! They'll help us survive!" So we eat it, and wow, does it taste amazing! "Especially with sugar," Brewer said, "our bodies send a signal to our brain that says, 'Remember what you're eating and where you found it.' We lay down this context-dependent memory and learn to repeat the process next time. See food, eat food, feel good, repeat. Trigger, behaviour, reward." Pretty soon, we have a brainstorm: maybe we should try eating a cupcake if we're mad or sad. After all, it gave us a buzz before. Why wouldn't it lift our spirits this time? Now there's a new trigger for the same cycle of behaviour and reward, and each time we repeat the behaviour, we reinforce the learning process. Pretty soon, numerous triggers push us toward that same reward. Hungry? A cupcake would sure hit the spot. Stressed out? Have a cupcake. Just got dumped? Cupcake time! Thus, Brewer explained, the brain mechanisms originally designed to help us survive wind up getting us hooked on habits that are bad for us. We get addicted to avoiding discomfort of any sort.

The same is true even when the substance doesn't induce pleasure so much as numbness. During his time at Yale, Brewer worked at a veterans' hospital as an outpatient psychiatrist specializing in addiction. Many patients hooked on opioids landed in his office at the behest of concerned primary care providers or family members. Working with these patients, Brewer was able to help them identify what he calls their "habit loop": the trigger-behaviour-reward pattern that was feeding their addiction. Let's say a soldier had a flashback of a traumatic event in Vietnam (trigger). To dull the pain, he got drunk (behaviour). The feeling of being drunk was better than the feeling of reliving his trauma (reward). In other words, the patients were using drugs not to feel good but to deaden the physical pain of their injuries and/or to lessen or dodge emotional pain of one kind or another. To avoid discomfort.

The vast majority of his patients, whether they were addicted to alcohol, drugs, or food, were reaping the same reward from their addictions: easing unpleasant thoughts, memories, or feelings. Pleasure wasn't the point. Rare was the hungover, broke patient who felt a three-day bender had been a wonderful experience. On the contrary, patients talked about wanting to avoid their feelings, and about succumbing to the pull of their cravings. It wasn't the language of happy people. It was the language of people who couldn't handle discomfort—who were in fact so driven to avoid it that they were doing something that felt even worse: destroying their own lives.

Brewer, still a devoted meditator, wondered whether mindfulness training—learning to pay attention to moment-to-moment experience without judging it—would help patients break the habit loops that led them to prefer self-destruction to discomfort. After all, Jon Kabat-Zinn's eight-week program, mindfulness-based stress reduction, or MBSR, was developed to help patients who were being treated for severe medical conditions and were in a lot of pain. Incorporating yoga, body scans, and meditation training, MBSR was designed to help people "find new ways to be in relationship to their pain," Kabat-Zinn has said, by training them to separate their actual bodily sensations from their emotions. "So when the thought arises, for instance, 'This is killing me!' instead of believing it, you investigate it. Is this killing me? No. Really, what you're doing is worrying."

Mindfulness, in this context, is all about focusing on discomfort—the very thing addicts seek to avoid. Rather than trying to dull the pain, addicts were urged to zero in on it. Brewer's initial forays into mindfulness were encouraging. One woman who came to his clinic was hooked on Entenmann's Louisiana Crunch Cakes: she craved them so much that she was powerless in their presence and would devour an entire cake in one sitting. We're talking almost twenty-five hundred calories in one go. Brewer taught her to pay attention during a binge, focusing on each mouthful and the bodily sensation she was experiencing, and she began to realize that the cakes didn't taste as good as she'd imagined, and that afterward she felt nauseated and guilty. Her "comfort food" was actually making her feel incredibly uncomfortable. Gradually she became disenchanted with eating cake, lost her cravings, and started to realize that fruit and other natural foods tasted better *and* made her feel better.

In 2008, by which point he'd become medical director of the Yale Therapeutic Neuroscience Clinic, Brewer decided to see whether mindfulness would work with the mother of all addictions: smoking. Because it's relatively inexpensive, doesn't impair cognitive functioning, and is tolerated more than, say, heroin use, smoking is one of the hardest habits to break.

For the study, Brewer and his colleagues recruited smokers who'd tried but failed to quit an average of six times, and they randomly assigned them to one of two groups. The smokers in the first group received mindfulness training and learned to meditate; those in the second group, guided by a psychologist, completed the American Lung Association's "Freedom from Smoking" program, considered the gold-standard treatment. Both groups met twice a week for four weeks, regularly switching rooms to ensure there was no chance the physical environment was influencing the outcome. At the end of the month, participants blew into a contraption that measured carbon monoxide to see if they'd quit smoking; if they had, they would exhale significantly less CO.

Those in the mindfulness group learned to identify their habit loops and map their triggers in the first class so they'd understand how they were actually reinforcing their habit with each cigarette they smoked. Instead of being sent home with the admonition not to smoke, they were advised to pay close attention to their triggers and the feelings and physical sensations they experienced when they did smoke. Brewer's job was to get them to see that in fact their "comfort" was creating even more discomfort, and his starting point was always the same: "I rubbed their face in the shit." In other words, he told them to really focus on what that cake tasted like, how it felt to shovel it into their mouths, what was going on in their stomachs, what sensations they were experiencing.

At the second class, three days later, many people reported that they'd found themselves smoking out of boredom. One man who'd realized he was smoking mainly out of habit or to mask the bitter taste of his coffee had cut down from thirty cigarettes to ten in two days. Instead of reaching for a cigarette, he'd started brushing his teeth. A lot of the participants couldn't believe that simply paying close attention had changed the whole experience of smoking. It tasted awful! One smoker remarked that it "smells like stinky cheese and tastes like chemicals. YUCK." While she clearly already had the cognitive awareness that smoking was bad for her (or she would not have joined the program), by becoming curious and attentive *when* she smoked, she came to realize that cigarettes, as Brewer bluntly put it, taste "shitty." Through mindfulness, she'd moved "from knowing in her head that smoking was bad for her to knowing it in her bones." The visceral distaste she'd developed for her habit almost overnight prompted her to become disenchanted with it—and its spell over her was broken. Nobody had to force her to quit.

The notion of force is important, Brewer explained, because many traditional smoking-cessation programs count on cognition, which resides in our prefrontal cortex (PFC). While that area of the brain understands perfectly well that we shouldn't smoke—and performs yeoman service trying to stop us from engaging in behaviour that we know is bad for us—the PFC is also the first part of the brain to "go offline" the minute we get stressed. When it does, intellectual reasoning goes out the window—and so does all the wonderful education you've had about why smoking is such a bad idea. System one is our reward system and system two is our PFC, Brewer told me. And "system one totally kicks system two's ass all the time." System two just isn't as strong or as well developed, he continued. "The cortex is the youngest—it's the newest kid on the block—so it can't compete with its evolutionarily much older, and not necessarily wiser, brothers and sisters in the brain." In other words, reward-based learning trumps earnest anti-smoking brochures every time.

Therefore, learning how to become disenchanted wit[h] our habit is crucial, Brewer believes. For change to happen, it must happen not intellectually but at the system one level, of trigger-behaviour-reward. The paradox is that to do that, we mustn't *resist* our cravings but actually turn toward them and focus on the discomfort we're experiencing. Mindfulness helps break down the cravings into transient bodily sensations. "I'm going to go crazy if I can't have a cigarette" becomes "I notice a feeling of tightness, tension, restlessness." If we focus on the "bite-sized pieces of experiences," Brewer explains, it's possible to manage moment to moment and "surf" a craving so we're not "clobbered by this huge, scary" wave. We don't have to rely on cognition to stop ourselves from reaching for a smoke because that wave gets weaker and weaker the harder we stare at it, until it loses the power to drag us under. And every time we don't get dragged under, that trigger loses its potency. We rewire our brain's response to the thing we crave.

When he began the smoking-cessation study, Brewer was hoping that mindfulness training would prove to be at least as effective as the American Lung Association's program. It turned out to be *twice* as effective at helping people quit. Furthermore, at the seventeen-week follow-up, more people in the mindfulness group remained abstinent, while those in the other group had lost ground, resulting in a five-fold difference between the two.

The take-away is that *attending to and learning to tolerate discomfort lessens its power and increases the likelihood of positive change.* Patients usually wind up in Brewer's clinic because the ways they've found to mask or avoid discomfort just aren't

working for them anymore. "Pain activates change," he said. What patients thought was a source of comfort, whether it's cake or crack, just isn't doing the trick any longer. It's a matter, he explained, of "recalibrating their systems [so they can see] that there are more rewarding rewards—such as peace and joy."

There are clear benefits to learning how to withstand discomfort. We'll be more comfortable with change and better able to innovate. We'll be less likely to develop self-destructive habits. And we'll be better able to cope with the stress and anxiety of everyday life.

Counterintuitive? Maybe. But another proponent of mindfulness says that letting yourself experience discomfort, without judging it, disperses tension and allows you to live in the moment. Andy Puddicombe is the co-founder of Headspace, an app that teaches guided meditation. Launched in 2012, the app has many high-profile fans, including Arianna Huffington, Richard Branson, and the Seattle Seahawks, as well as, Puddicombe says, more than six million other users. When you download the Headspace app to your phone or computer, you can opt for a garden-variety ten-minute guided meditation or choose from a menu of more targeted options designed to improve your health, performance, or relationships. Puddicombe's voice guides you through the hundreds of hours of instruction, with each session starting with a "checking-in routine": take a few deep breaths, notice any background noise, and become aware of your physical sensations. Gradually, he draws your attention to your breath, gently asking you to count your breath in and out, in sets of ten.

People who have never meditated before often ask what you have to do. The aim, however, is "not to *do* anything," said Puddicombe, who has a shaved head, a Bristol accent, a self-deprecating manner, and the look of a man for whom fitness has been a lifelong pursuit. "To some extent, we're simply sitting down, watching the mind. We're not projecting onto the mind. We're not saying, 'This thought is good' and 'That thought is bad.' 'I like this feeling,' 'I don't like that feeling.' As long as we're doing that, we're just participating in the same kind of inner chatter that's gone on our entire life." That inner chatter is what makes us anxious and distracted, unable to live fully in the moment. "If we can learn how to sit there and witness thoughts and feelings without participating in that, and instead simply see them—'Oh look, a thought,' 'Oh look, a feeling'—all of a sudden the mind starts to feel much more content and at ease. And then we're not involved in strategies on how to get rid of anger or how to banish fear. Instead it's kind of like, 'Oh look, there's fear.' And we're no longer fearful of it."

In other words, like Brewer, Puddicombe believes that focusing on discomfort has the paradoxical effect of making us more comfortable—and better at facing change and other challenges. Our brains tend to "add layers and layers of difficulty" when we're already facing a challenge, according to Puddicombe. "We grab hold of it internally, and we think about it," but not in a way that's productive or helpful. Instead, worries, fears, and random thoughts "swim around and around in our minds," and we get caught up in them, "wishing it wasn't like this, or wishing something had happened differently in the past, or looking forward and hoping it's going to end up a certain way, or fearing it might end up a certain way."

Meditation also gives the younger part of our brains—system two—a hand. It has been shown that meditation pumps up the volume on the grey matter and synapses in the prefrontal cortex—which, in addition to governing our pain responses, helps keep us grounded and focused. A buff prefrontal cortex is important because the amygdala (an almond-shaped mass of grey matter located deep in the medial brain's temporal lobe, and part of the older, more primitive system one) is wired to protect us from harm. Essentially an early warning system, the amygdala is constantly scanning for and interpreting subliminal danger signs. The minute it senses we're in harm's way, it triggers a visceral reaction—fear, for instance—to tell us this might be a good time to cut and run.

Experienced meditators are good at not reacting to such triggers. They've developed the ability to disable the negative emotions and fears surrounding a reflexive thought—"I'm going to flunk the exam," say, or "My new boss is out to get me." Instead of being derailed by discomfort, these people explore the thought neutrally, separating fact from fear.

Mindfulness meditation has been shown to reduce anxiety significantly, even for those who've been diagnosed with clinical anxiety disorders. The effects are not just short term. As much as three years after an eight-week mindfulness course, according to one study, subjects had maintained significant gains. Most of them had stuck with mindfulness meditation and continued to practise it, because the benefits were so clear.

Puddicombe was eleven years old when he was introduced to meditation by his mother, an acupuncturist—though when he agreed to accompany her to a class, he was under the impression that they were going to kung fu training. He recovered from his disappointment, taking classes on and off as a kid, but by the time he started university in Leicester, England, he was more interested in partying than inner calm. Then one evening when he was leaving a party with a group of friends, a drunk driver barrelled headlong into the crowd, killing several people and critically injuring many others. Shortly thereafter, Puddicombe's stepsister died in a cycling accident. Plunged into an existential crisis by these tragedies, he felt that everything he'd once cared about was meaningless. It was, he said, "a sort of early midlife crisis."

One day in his dorm room, he had a profoundly moving spiritual experience, the essence of which he still struggles to articulate. All he can say is that when it ended, he wanted to become a Buddhist monk. His friends and family thought he'd lost his mind. The university urged him to take some Prozac and finish his degree. Instead, he boarded a plane to India, took a vow of celibacy, and spent the next ten years cloistered in Buddhist monasteries there and in Nepal and Burma.

When he first began practising meditation, Puddicombe told me, his mind really appreciated the chance to slow down, and for the first couple of months he went through a honeymoon period. Then it got hard. Really hard. "There's no TV, no books, no people you're speaking to. You're just kind of stuck with these thoughts—the thoughts and feelings that you spent a lifetime trying to get away from. And then all of a sudden, you realize, 'Oh, no! Now I have to sit here with them *every day*.' So that's challenging." At one point, Puddicombe was having four-hour meditation sessions four times a day. When he was in Burma, his teacher, a monk, spoke no English, and since Puddicombe didn't speak Burmese, their daily tête-à-têtes consisted of exchanging smiles and the occasional knowing nod. And of course, he spent years sitting still, alone with his thoughts. But the more he surrendered to the process and pushed himself to focus on his thoughts, the more comfortable he became.

At thirty, no longer searching for the meaning of life and newly open to the idea of other avenues, he decided that rather than sequestering himself as a monk, he might be ready to head home. But first, Puddicombe detoured to Russia to train with the Moscow State Circus. His radically different life had begun. Once back in London, he continued to study circus arts during the day; at night, he worked on developing corporate-friendly meditation courses. Who, he figured, could benefit more from mindfulness training than stressed-out executives? In the final year of his circus training, he was balancing a German girl on his head for a magazine photo shoot when his neck gave out and he wound up in hospital. In rehab he met a sports physiotherapist named Lucinda, whom he later married. And his time in rehab was fulfilling in another way: he was hired by the clinic to teach meditation, and his new career began.

His timing was impeccable. Meditation had already gone mainstream, in part thanks to myriad studies showing that the practice helps combat depression, high blood pressure, and insomnia. From his years as a monk, Puddicombe knew it could do even more. If you devote enough time to meditating, there's growing evidence that it can actually change your brain. Literally. The Dalai Lama himself helped recruit Tibetan Buddhist monks for brain-imaging studies, the findings of which, bolstered by other research, suggest that if you put in the hours, meditation will alter the structure and function of your brain in positive ways.

Although Puddicombe was well aware of the health benefits of meditation, what really inspired him to spread the word was its ability to help people focus their attention and more fully inhabit their lives. He often mentions a 2010 Harvard study showing that people spend about 47 percent of their time thinking about something other than the present moment. "Almost half of our life has been lost in thought," he told me, sounding genuinely horrified, as though the idea had just occurred to him. "As if we were just running on autopilot!"

"A human mind is a wandering mind, and a wandering mind is an unhappy mind," wrote the two psychologists who authored the Harvard study. "The ability to think about what is not happening is a cognitive achievement that comes at an emotional cost." Since we're not on earth for very long, Puddicombe thinks it's tragic that we spend almost half our lives too distracted to really live them. At the end of his own life, he doesn't want to realize that he was so "caught up in his own stuff" that he didn't stop to notice his wife's smile or his child's first steps or the person at work who needed his help. "I just feel those are the things we have the opportunity and potential to notice . . . to be awake in our life, but so often we are just on autopilot." While autopilot may seem like a safe and comfortable state, the Harvard study demonstrated that it isn't.

Marketing Headspace as a "gym membership for the mind," Puddicombe has become something of a guru to the stars—albeit an unusually down-to-earth one. For him, meditation is not just big business but a way of life, and one that has helped him weather both the ups of digital celebrity and the downs of serious illness. In 2013, shortly after he and Lucinda married and moved from London to LA, where the Headspace headquarters had relocated, he discovered he had an aggressive form of testicular cancer. The news was shocking—even a former monk doesn't take a cancer diagnosis in stride—and without meditation, Puddicombe says, "I think I would've felt helpless, probably quite isolated."

Really digging into his fear about what might happen, however, gave him a sense of comfort with his own discomfort and made it possible to "rest" in uncertainty—"not knowing, not thinking it was unfair, and not wishing it was different. Not having this idea of having to fight something or battle something. And not looking too far forward and thinking, 'Well, what's going to happen when the scan results come back, and what's going to happen when they operate?'" Instead, he was able to quiet his mind and stay in the present moment, thinking, "Right now, all that is happening is I'm sitting here. Or I'm standing here. Or I'm drinking a cup of tea."

Mindfulness, Puddicombe serenely declared in his 2012 TED Talk, "shows us that things are not always as they appear. We can't change every little thing that happens to us in life. But we can change the way we experience it." We can reframe our experiences, in other words, so that discomfort does not derail us or push us toward self-destructive behaviours, and instead becomes a sign to be studied, a compass that points us in the direction we need to go, whether that's away from a bad habit or toward growth, change, and a new way of life.

CHAPTER TWO

Reframing Discomfort

The pain started on day seven, about three miles into Ray Zahab's run. Pain that was noticeable and sharp, emanating from his left heel. At first he thought it was a blister, and his plan was to do what he always did in such circumstances: ignore it and keep running. An ultra-marathoner accustomed to rough terrain and gruelling conditions, Zahab is no stranger to discomfort; in 2007, he ran almost five thousand miles across the Sahara Desert in 111 days without a single day of rest.

For most mortals, completing just one marathon is a major achievement. But ultra-marathoners typically run for days on end, completing multiple marathons back to back—sometimes two in a single day—in the least hospitable places on the planet. From the Arctic to the Amazon jungle, these athletes test the outer reaches of their endurance. And when they hit a limit—physical or psychological—the very best of them, like Zahab, strain to push past it.

But this day in 2011 was different. What began as a routine irritation soon felt like a knife being jabbed into his heel. Coming to a stop, Zahab flopped down on the rocky sand and gently removed his sky-blue running shoe, then peeled off a sock slick with sweat. What he saw shocked him. It was a blister—or rather, it had once been a blister. Now the palm-sized disc of skin had peeled back, revealing a deep, angry-looking wound. Clearly infected, it must have been festering for days, unnoticed among the other aches and pains, causing a fever that had also gone unnoticed in the withering heat. Zahab was crossing Chile's rocky, barren, and vast Atacama Desert, one of the driest spots on earth, a place where temperatures can climb to over 120°F. The sun's rays mercilessly burned through his long-sleeved shirt and his gloves; on the salt flats, his exposed skin was quickly abraded by the wind. And the terrain—in the Atacama, stony ground gives way to small sliding rocks (picture running on broken glass) and soft sand—was as unforgiving as the climate. Nevertheless, Zahab had planned to cover about forty-five miles a day while carrying all the supplies he needed in a heavy backpack. Now, less than halfway through his fourteen-day run, he could barely stand. This wasn't just a blister. It was a disaster.

The day had started well. Roads are few and far between in that part of the world, but Zahab's three support team members had discovered an abandoned railway track he could run along. The silty sand was still tough going, but the track was relatively flat—a distinct improvement over the hard rocks of previous days. His plan had been to run for more than thirty miles before reconnecting with the team, which was inching across the desert in a four-by-four. Clearly, it was time for plan B. He used his satellite phone to call for help, and luckily, since it was early in the day, the team wasn't too far off. Even so, in the forty minutes it took for them to arrive, Zahab's foot ballooned alarmingly. He was soon feeling something worse than searing pain: he was close to despair. If there is one impulse an ultra-marathoner knows to obey, it is to keep moving. Even when reduced to crawling up a sand dune or down a hillside, you *move*. But moving was now out of the question.

By that stage of his career, the forty-two-year-old Zahab was well known; a film about his Sahara run had been narrated by Matt Damon. Thousands of fans and supporters were tracking his progress via web-link, and he connected with them nightly after pitching his tent. He hated the thought of letting them down. His gloom about the fate of the expedition was echoed by the words of the doctor his team tracked down in an old mining town after several hours of driving through the desert. The doctor checked out Zahab's injury and announced, "You won't be able to run on this foot."

Oddly enough, that declaration—so definitive, and so negative—raised Zahab's spirits. It sounded like a challenge—a dare, almost. A man who's spent months of his life running across deserts doesn't shy away from a dare. "It snapped me out of my funk," Zahab remembers, laughing slightly. "I thought, 'There's got to be a way.'" But before he could even verbalize this, one team member, Bob Cox, looked at him and said, "You're not going to stop, are you?" The answer was a resounding no. Zahab felt more determined than ever. Hadn't he run across the Sahara? What was another week of discomfort compared to that? Pain was just another challenge to be overcome, and the key was not only to ignore the physical pain but also to pay no attention to the naysayers—especially the naysaying voice in his own head, which was so much louder and more damning than anyone else's.

He wasn't going to be an idiot, though. No point jeopardizing his running career. He'd give it one more day, he decided, then reassess the situation. "I knew I could find a way to run through the pain—I'd run through worse than this," he explains. "But I also knew that if at the end of the next day I looked at the infection and it looked even worse, it would break me mentally." In other words, if his mind-over-matter approach failed to yield results, he would lose faith in himself in some fundamental way that would be more damaging than any bodily injury he'd sustained.

And so the next morning, Zahab cut his shoe open to accommodate his swollen foot, wrapped tape around the shoe to hold it together, and downed some antibiotics. Then he set off. "It was a whole other level of pain," he recalls. "First it just hurts. *Really* hurts. Then it almost goes to a place that feels itchy—almost like it's tickling you. But it is still incredibly painful." Zahab used every mental trick he could think of to overcome the pain, distracting himself by ruminating about various problems in his life, thinking of his family back home in Ottawa, and picturing his route broken into small increments so it didn't seem overwhelming.

After twenty-eight miles, he called it a day, satisfied enough with his progress. And then, heart beating fast, he sat down to take off his shoe. Miraculously, the infection looked pretty much as it had that morning—hours of bearing weight and being slammed into the ground with every stride hadn't seemed to make it any worse. That was that. "I'm going to crank it," Zahab told his team.

And he did. He went back to running forty-five miles in a day—not every day, but often enough to keep him on schedule. By the end of the expedition, his foot had almost completely healed, with just one tiny bandage marking the spot on the sole of his foot. He'd literally outrun the infection.

Zahab is fit and lean, with the narrow torso stripped of body fat you'd expect of a marathoner. But he's not superhuman. Nor is he one of those lifelong athletes who has been pushing himself (or has been pushed by others) from the time he could walk. As a kid, he hated gym class; throughout his twenties, he drank and smoked and didn't run at all. So how did he manage to pull off a feat of superhuman endurance even when his foot was a mess?

The answer, it seems, was all in his head: he'd found a way to make discomfort work for him.

How we respond to discomfort—or its ugly stepsister, pain—is no straightforward matter. Pain was once thought to be quite linear: a nerve ending in the body is injured or damaged, and it sends a signal to the brain to perceive pain. The thinking was that pain was like a telephone line that ran from the injury straight down the line to the brain, where it rang a bell: ouch!

But it turns out that pain is far from linear. We now know that it's what scientists refer to as "plastic," with a range of influences between the trigger event and the moment of perception in the brain. Because the brain itself is also plastic, pain can beget more pain—"like a computer rewiring its own software," according to landmark research on the subject.

Rather than a simple sensory experience, pain is a phenomenon governed by several different parts of our brains. The prefrontal cortex (PFC, or system two) assigns meaning to the pain, while the limbic system (our primitive brain, or system one), which governs emotions, dictates how we feel about it and can be responsible for triggering fear (which, in turn, exacerbates the severity of the pain we experience).

Let's say you fall and sprain your wrist. You might expect a fairly rapid signal to the brain—something along the lines of "God almighty, that hurts!" But in a minute fraction of a blink of an eye, different parts of your brain weigh in, deciding what the pain *means*. If you sprained your wrist falling out of a tree trying to help your three-year-old, who's dangling perilously from a high branch, your PFC might well send a strong signal to ignore the pain—there is still a toddler to be saved. If you fell in the last few steps of an Olympic event, your limbic system might telegraph powerful feelings of disappointment and shame, thereby intensifying the pain. Or maybe it will position the pain as the cost of being a good sport, thereby reducing its debilitating impact. In other words, your experience of "pain" is influenced by your current state of mind, your prior experiences, and your future expectations. All of which means we can't possibly appreciate the pain of someone else's blistered foot. Not only is another person's brain uniquely constructed, but the variables that add up to that individual's experience of pain are unique and ever-changing.

When we see trained athletes enduring major league pain, it's natural to assume they must not feel it the way we ordinary mortals do. But that's not what's going on. In 2012, the journal *Pain* published a meta-analysis of several studies—fifteen in all—that compared athletes' pain tolerance to that of active but not hugely athletic people. They concluded that the pain threshold—or the point where individuals *felt* the pain—was the same for both groups, but athletes were consistently better able to tolerate it.

Why? Some researchers think athletes have a higher pain tolerance simply because they're more highly motivated to ignore the fact that they're hurting. The sense of accomplishment that comes with winning a medal or beating a record can serve as a formidable incentive. One study revealed that some female British rowers actually considered injury to be a desirable aspect of their sport—which may speak to the fact that athletes equate pain with progress, or to the messages they receive from coaches and the culture at large that it's heroic to tolerate pain. But the bottom line is that for athletes, pain can be a signal they are on the right track. They've reframed how they perceive the pain as a positive, not a negative.

For athletes, mental and physical discomfort is also a known occupational hazard, not a shocking surprise. Whether they're going for gold or simply trying to get through a tough training session, athletes accept discomfort as a part of sport. Viewing it as necessary and inevitable not only helps them tolerate it, but also lessens its power over them. And many athletes take comfort in the knowledge that they're not alone. Knowing that teammates and opponents are persevering through pain spurs them on to tolerate their own with more equanimity.

Studies have found that athletes sometimes gain control over their pain by learning to differentiate between the grades of pain they're feeling, and by assigning different values to different degrees of pain. Once they know what the pain signals, in other words, they know how to react to it. Routine performance pain—generally muscular in origin, short term, brought on by training, and therefore under their control—doesn't deter serious athletes. On the contrary, it signals, "I'm doing well." (Athletes are able to parse their pain in remarkably nuanced ways. In one study, dancers' descriptors for their "good" pain ranged from "achy," "nice," "throbbing," "nagging," and "niggling" to a "crunchy heat" and the feeling of "ants walking on you.") The satisfaction they derive from knowing they're pushing hard enough to feel the pain motivates them to push even harder. Thus performance pain not only becomes their coach—it becomes their cheerleader too.

Injury pain is another matter: it rings alarm bells and announces, "I'd better stop now or I'm in trouble," or "I'd better get this checked out." This brand of pain can manifest as a warning twinge or an acutely intense or numb sensation—the latter is a greater cause for alarm because it suggests the body has been pushed beyond its limits. Whatever its degree, though, injury pain tells the athlete the one thing he or she most dreads hearing—namely, "The situation is beyond your control."

When it comes to discomfort, control is everything for an athlete. Pain management research shows that the degree to which athletes believe they have mastery over their pain plays a big role in how they respond to it. The minute they start catastrophizing—"I'll never make the Olympic team now!"—their ability to tolerate pain goes into free fall: their brains are essentially hijacked by cascading emotions like anxiety and fear, muscle tension ramps up, and the neural systems that heighten pain sensitivity go into overdrive. Next comes the precipitous plunge in confidence around their ability to cope with the pain.

If, however, they view pain as a challenge to be overcome, they're much better able to talk themselves through it—which is why, when pain hits, the first thing serious athletes tend to do is ask themselves a series of questions: Is this a dangerous pain? If so, just how bad are we talking? Is it the kind that's going to sideline me for a season, or the kind I can harness to light a fire under myself?

The message is clear: believing you can control your feelings of discomfort heightens your ability to cope. And you don't have to be a high-performance athlete for that to be true. If you interpret your discomfort as threatening—or obsess about it instead of focusing on your goal—you will suddenly be living in a world of pain. If, however, you view discomfort as necessary, inevitable, and even an encouraging sign that you're on the right track, it just might be your best friend.

• • • •

Ray Zahab isn't afraid to exert energy—physical or mental. On one occasion, he ran almost two miles to meet me and arrived breathing naturally. He hadn't even broken a sweat. Looking at him today, it's easy to imagine why he became an ultra-marathon runner: he was born for the part.

Only, he insists, he wasn't. Though reasonably fit, he was the kid who usually got picked last for the team. He wasn't a jock, or even much of an athlete, and he dreaded gym class. He grew up on a small horse farm near Ottawa, Ontario, but he was born with a congenital spine condition, so he'd been warned to avoid horseback riding and running. Every day began with pain for Zahab: when he woke up, his back was stiff and sometimes one of his legs was completely numb. But his parents were matter-of-fact about his condition—no point dwelling on it—and so was he. "I did physio, and that was it. It didn't occur to me that everyone didn't wake up in pain. It was just no big deal."

As a teenager, he was unmotivated, insecure, and possibly more than a little bit depressed from time to time. But he hid his unhappiness behind a gregarious personality, and even those who knew him well wouldn't have guessed that his internal monologue was resolutely negative. "You'll never amount to much" sums up the refrain.

On the outside, Ray Zahab was laid-back, the kind of guy who drinks and smokes and takes it easy whenever he has the chance. He also had a bit of a rebellious streak: despite having been told not to ride, he was an expert horseman by his late twenties, competing in quarter horse reining events and training other people's horses as well. Clearly, he was capable of pushing himself, but he preferred to drift along, letting life happen to him rather than setting goals and seeking to meet them. He didn't think of himself as a dogged fighter or a striver or even a guy with a sense of direction. In his own mind, he was always lesser than.

One day when he was in his early thirties, though, he woke up with a vague but compelling itch: *I want to be happier*. Life was passing him by. He knew he had to do something, but he had no idea what it should be. He'd always been close to his younger brother, John, however, so when he began to ask himself how to make positive changes in his life, it felt natural to look to John, a successful and driven personal trainer, the kind of guy who radiated both a sense of purpose and a joie de vivre. What had John done differently? What motivated him? And what lessons could Ray apply to his own life? Asking those questions—even subconsciously—was the easy part. Being open to the answers, which were slow to come and sometimes buried deep within an experience, took longer. Acting on them to change course would take longer still.

Zahab began his search for a happier life by emulating his brother, who seemed to have it all figured out. John loved outdoor pursuits: climbing, hiking, and cycling. And so Zahab began to join him, at first sporadically, but eventually in a more committed way. He even started his own personal training business, building on the knowledge he'd developed helping his riding clients strengthen their core abdominal muscles. But in some ways, his copycat approach remained haphazard. The old Ray—not sure he could push himself very far, or even if he wanted to—was still around, and so he didn't commit to his business in the same way his brother did. He hung back, doubting himself. What if he tried harder but failed? It was the same fear that had dogged him in high school gym class.

His epiphany came at the top of an icy rock face. He and John had decided to scale a local mountain, relying on ice picks and clips and climbing skills the elder brother frankly did not possess. Physically stressed and slightly terrified, Ray was nevertheless determined to keep up with John. He's still not quite sure how he pulled it off, but getting to the top of that mountain changed something inside Ray Zahab. By venturing far outside his comfort zone, he'd accomplished something that previously had not seemed possible. He had pushed past his fears, past the uncertainty about his own ability, and past his own considerable discomfort while actually climbing. The mental triumph felt even more exhilarating than the physical experience of standing on the summit, looking down at the forbidding pile of ice and rock. His big question—can I change myself and become stronger and better?—had been answered.

Before he'd even completed his descent, he knew he wanted to feel that same sense of victory again. And he also knew he couldn't get it by attempting something easy. The high he'd felt on the mountaintop was inextricably linked to the possibility of failure. He couldn't get the high unless he was willing to experience the discomfort of risk, the discomfort of stretching beyond what he knew for sure he could achieve.

He decided to take up a sport, but not in a half-hearted, Sunday afternoon way. In what would come to characterize his style—impulsively biting off more than he might be able to chew—he agreed to compete in a serious cycling race long before any sensible person would have considered it advisable. He didn't, it seems, want anything to feel easy or comfortable. Whether he did so consciously or not, he chose the hardest path, the one where success would be most difficult—and as he'd learned on that mountain, also most meaningful to him.

He didn't give up his personal training business, but cycling became his focus. Zahab enjoyed training and quickly found support from fellow cyclists and potential sponsors (the latter essential to any athlete taking on global competitions). Along the way, he also learned that his muscles were slightly different from other people's: rather than powering through the "burn" (that "good pain" associated with building new muscle mass), he didn't even feel it. He simply pedalled until his leg muscles gave out.

It wasn't long before the sense of self-mastery he was experiencing on a bike influenced other areas of his life too. Like most smokers, Zahab had been trying to quit for years. But New Year's Eve 2000 provided a good symbolic break, he decided, and he smoked his last cigarette just before midnight. When he woke up the next day with the inevitable and intense craving for another, he started the new millennium by putting a pack of cigarettes in the hip pocket of his jeans—and then ignored them. "I needed to show myself [that] I was bigger than the cigarettes; that I was in control, not them." He never smoked again. Good thing too, because his athletic career was about to get even more serious.

A serendipitous moment with a travel magazine in his chiropractor's waiting room turned his life in a new direction in late 2003. Actually, that chiropractor's office changed his life in other ways too—it was where he first spotted Kathy, the woman who is now his wife. But what struck him on this particular day was an article about the Yukon Arctic Ultra, a hundred-mile multi-day marathon taking place just a few months later. Runners would have to carry every piece of gear and food they'd need for the six or seven days it took them to go the distance; the cold would be intense, and running on snow and ice would be incredibly difficult. Meanwhile, storms, dehydration, and disorientation all presented real dangers. Zahab was intrigued by the challenge, but also curious: "I wanted to know what kind of people would do something like this." The obvious answer—"people like me"—eluded him, because he wasn't yet in the habit of thinking of himself as someone who was galvanized by a challenge. The tougher and less comfortable, the better.

Training for an entirely new sport was intense. All the work he'd done to become a cyclist was next to useless when it came to running long distances on the trails near his home. His first time out, he aimed to complete a five-mile loop but had to stop at the four-mile mark. Thanks to the cycling, he had cardiovascular health on his side, but the pounding on pavement was new: his hips, legs, and feet hurt like hell. Nevertheless, he says he "felt like a runner" almost right away. Gradually increasing the length of his runs—and running on snowy paths to try to mimic the terrain he would face in the Yukon—he built up to twenty miles within a month. Next he had to learn to run while pulling a sled loaded down with cold-weather gear.

Running in a place like the Yukon isn't just about stamina—there's also thin ice, frostbite, and hypothermia to worry about. Some runners camp at night in order to rest; others simply stop in their tracks, sleep for a few minutes, then carry on. Zahab decided to try the latter approach, though it meant running in the dark—which was treacherous—guided by a small headlamp.

Race day. At the first mandatory checkpoint, twenty-five miles in, runners had to prove they were carrying certain gear and knew how to use it. Zahab did all that, then was astonished to discover he was in second place. Alternately running and walking, and occasionally crawling, he finished the race exhausted and dehydrated. And in first place.

Along the way he'd discovered what *really* testing his limits felt like. Initially, his goal was simply to complete the run, but as the days wore on, he realized he hadn't dreamt big enough. Focusing on finishing wasn't enough; he started to focus on ways he could push himself even harder each day of the race. "There is the grand journey to the finish," he explains, but he'd realized that "there are smaller components too"—little challenges you can focus on each hour to keep pushing yourself toward success. "I learned not to be afraid of being afraid," he says. Self-doubt—can I actually do this?—became a feeling he not only tolerated but actually began to crave.

Discomfort—physical, emotional, mental—was, he realized, not a signal to hesitate or stop. It was a signal to dig deeper. Try harder. Discomfort was an essential ingredient of his success because when he reframed the discomfort, he used it to push himself to go farther, faster. "[We] human beings are all genius at talking ourselves out of doing something," Zahab asserts, but he firmly believes that anyone can flip the mental switch and embrace the discomfort that comes with trying new things and testing limits. These days, he makes a living helping others do just that. When he's not racing to raise money and awareness for causes like protecting the water supply in sub-Saharan Africa, he's taking groups of teenagers on extreme expeditions, testing them physically while letting them design an experiential curriculum that matches what they need to learn. "I wish I could go back and tell my seventeen-year-old self what I know I'm capable of," he says. "Instead, I try to help other seventeen-year-olds figure that out now." To do that, he created a company called impossible2Possible, or i2P.

In a typical i2P expedition, five seventeen- to twenty-one-year-olds from all over the world run the equivalent of a marathon a day for six straight days in, say, the Amazon jungle. Meanwhile, they study the region's ecosystem, sharing their findings via webcam with their fellow students back home. Along the way, they experience failure, injury, discouragement, fear—and have to reframe the experience and their responses. Like their mentor, they often make transformative discoveries—although not necessarily the ones they'd been expecting to make.

Hannah Elkington had only ever completed a short community run when she marched up to Ray Zahab after a presentation and told him she was signing on to spend five days running thirty miles a day across Botswana's Kalahari Desert. As a competitive wrestler and rugby player who planned on a career in policing, Elkington was no shrinking violet. She was disciplined and she had experience hitting targets. For the next five months she trained hard, waking up at 4:00 a.m. each day to fit in a long run before heading off to school. To bolster her confidence, she also spoke regularly with the other members of the expedition.

Day one in the Kalahari was incredibly hard, but when the runners passed the twenty-two-mile mark, they high-fived each other. "We were beginning to realize this was doable," Elkington remembers. Day two, however, didn't turn out so well for her. At the sixteen-mile mark, she collapsed: dove headfirst into the sand and drifted in and out of consciousness for the next ten minutes. It turned out the malaria pills she'd taken had increased her sensitivity to heatstroke, and her body had stopped processing water and electrolytes. The doctor said she wouldn't be running any time soon. On day three she was still lethargic—and emotionally crushed. As she and Zahab talked through her sadness about not reaching her goal, her perspective gradually began to shift. Her goal had always been to finish the race and not let the team down. She couldn't run it, but what if she finished it another way? That day, joined by another runner who'd also been felled by heatstroke, she walked twenty miles behind the group. To avoid overtaxing their bodies, the two walked for a stretch, hitched a ride on the support truck long enough to cool down, then hopped off and kept walking—remaining true to the ultra-marathoner's credo: *Don't worry about your speed. Just keep moving.*

Before embarking on the expedition, Elkington had always assumed that the life-altering experience would be running the race to the end—just as it had been for Ray

Zahab. But when that goal was no longer within reach, she found a way to reframe her disappointment and focus on an objective that would still allow her to feel a sense of achievement and pride. And that ability to reframe disappointment and keep moving forward turned out to be the experience that has stayed with her. It was a powerful lesson for a seventeen-year-old to learn.

Zahab's desire to impart such lessons—and his willingness to live at the extreme edge of his own comfort zone—had a profound impact on the way he went about building i2P. He wanted a full-sponsorship company so that students and their schools would get everything for free. "Kathy and I invested everything we had in it," he says, but pursuing other funding was daunting. "You take this idea to people, and you think it's a great idea, yet you're waiting for *them* to tell you if it's a good idea or not." His first big potential sponsor shot down his pitch, saying, "We love it, but we don't really understand the business model." Zahab was disappointed, but instead of thinking, *I've never run a business, so they're probably right*, he declared, "What the hell—I'm going for it. There may not be a business model here, but I know we can make it work." He had learned to reframe discomfort in other parts of his life too.

Zahab is not immune to self-doubt, and in fact he considers it an essential component of success, especially if it's the kind that "makes you lie awake at night worrying about everything that could go wrong." That sort of pervasive doubt is actually what helps him prepare to face a major challenge. The trick, he says, is to meet doubt head-on without being terrorized or paralyzed by it. You need to confront discomfort and view it as a helpful reminder to prepare properly, rather than a signal to quit.

When they're under stress, how do athletes like Zahab distract themselves from discomfort? Some, like long-distance runners, learn to shift gears between focus and distraction. On the one hand, they tune into their fatigue and pain in order to assess their performance and make decisions; on the other hand, they intentionally distract themselves from it by talking to a running mate, letting their minds drift off, or simply gazing at their surroundings. This "associative-dissociative strategy" allows them to monitor their pain without being laid low by it. This is how Zahab chugged through the desert hour after hour, the day after the doctor had ordered bedrest.

Some researchers theorize that the endorphins released by an athlete's body when he trains and performs act as a natural analgesic—the so-called runner's high. It's almost as if when the pain messengers travelling from the athlete's muscles to his brain arrive and knock on the door, the brain is so high on endorphins it barely hears the sound.

The other strategy that helps an athlete stay the course is remembering that eventually the pain will pass. (Knowing how long you'll have to put up with pain is a big determinant of how well you'll be able to tolerate it—not just in athletics but in the rest of life. In a lab study where subjects were exposed to a "noxious stimulus," those who had control over how long they'd have to endure it reported less pain than those who had to rely on an experimenter to call time.)

Despite the immediacy of pain, if an athlete focuses on the finish line—and trying to get there ASAP—it can make all the difference.

Bottom line: athletes don't get a free pass on pain. Like Ray Zahab, they just find ways to reframe how they think about it—by expecting it to happen, training their brains to view it as a positive thing when it does, and relying on their brains to pretend it's not really there. And the meaning we assign to pain matters. A lot. Consider the real-life experience of climber Aron Ralston, who found himself trapped by a rock slide in southeastern Utah for 127 hours (which became the title of the Hollywood film about him). Ralston worked frantically over three days to try to free his trapped arm, knowing his survival depended upon it. The idea of severing his arm occurred to him early on, but he decided that would be nothing more than a slow form of suicide. When it became apparent that death was certain if he didn't take action, however, he changed his mind and set about cutting off his arm to save his own life. The amazing part of Ralston's story isn't his willingness to do such a grisly and dramatic thing—the will to live has propelled many people to perform remarkable feats—it's that he later reported feeling no pain while he amputated his limb in a slow and meticulous fashion. Or rather, the pain he felt, which can only have been intense, was no more important a sensation than the many others he was experiencing. The magnitude of his task—saving himself from certain death—changed the way his brain perceived the pain he was experiencing.

From a climber making a life-or-death choice to an ultra-marathoner dragging a sled across the frozen north to a couch potato who's on a treadmill for the first time in years, understanding the "why" of pain is obviously important. Having a very clear goal, as Aron Ralston did, helps. Thinking of the pain as a work in progress, a means to an end—and, crucially, as something that will pass—also makes it more bearable.

When we reframe discomfort, either by associating it with something positive or by treating it as normal and even helpful, we engage the prefrontal cortex, our brain's CEO, which can act as a top-down neurological commander to close the gate on unbearable degrees of pain.

It isn't just physical pain that travels this neurological corridor of smoke and mirrors. Many forms of discomfort that are triggered when we try to change our circumstances or ourselves—financial uncertainty, fear of failure, and even our innate distaste for any ambiguity—are also influenced by various parts of our brain, as well as by our own experiences and perceptions. Like athletes learning to reframe the pain of training as beneficial, we can teach ourselves to get comfortable with the discomfort that goes along with change. The first step is simply to change our own minds about what discomfort means, and consider the possibility that it's a sign we're stretching and growing.

CHAPTER THREE

The Mindset Factor

If the nurse on the afternoon shift had worked a different floor or been off that particular day, his life might be very different, or so it seems to Jim Moss. In September 2009, he was bedridden in a San Jose, California, hospital, fighting to recover from an autoimmune disease that had robbed him of the ability to walk without assistance. But Moss hadn't parted with his dignity. Unwilling to "pee in a jar," as he put it, he'd asked for help one morning to get from the bed to the bathroom, ten feet away. Possibly impatient with just how long the journey was taking, the nurse he was leaning on remarked grimly, "You'd better get used to this. You're going to be like this for a long time." Nurses see everything and usually know a lot more about patients' daily struggles than doctors do, so her words carried the weight of authority. Her defeatist message rattled Moss, a professional athlete accustomed to pushing through various forms of physical discomfort to achieve his goals. When she left the room, he was overcome by the depressing feeling that life as he knew it was probably over.

But that afternoon, when he again summoned help to get to the bathroom, a different nurse turned up, and this one had a very different personality and outlook. He remembers that she was short and jolly, and as they paused for a breather during that brief journey across the room, she looked up at him cheerfully and said, "Don't worry, sweetheart—we'll get you back on your feet in no time." Her words also carried the weight of authority; she was as much an expert as the nurse he'd seen in the morning. Moss's feeling of despair abruptly vanished, to be replaced by buoyant optimism. How could thirteen words have such a profound emotional impact? "It's the semantics," Moss explains today. "*We* are going to get you back on your feet—we are both going to play an active part in doing this. And she said 'in no time.' 'No time' and 'a long time' can be the same [length of] time, depending on your frame of reference," but one sure sounded better. To Jim Moss, this was no throwaway remark—it was a game changer. There was hope for him after all. As it turned out, the resurgence of optimism kick-started by that nurse wound up setting him on a path that changed his life.

Almost accidentally, Moss had discovered what science has known for some time: that our frame of mind dramatically influences how we respond to our circumstances. A positive frame of mind makes a negative event easier to bear, and ups our level of satisfaction. The key is learning how to change your mindset, and again that cheerful nurse working the afternoon shift provided a key. Reframing discomfort can be as simple as casting it in the best possible light. "No time" and "a long time" could both mean one year—but the former definitely sounds more optimistic. Choosing the more positive way reframed discomfort for Jim Moss.

Today, fully recovered, Moss looks every inch the serious athlete. Six feet tall, with tousled dark hair and piercing green eyes, he has wholesome good looks and an easy manner. In the fall of 2009, he was a professional lacrosse player for the Colorado Mammoth and a former professional hockey player for the Huntington Blizzard of the East Coast Hockey League. On September 19, after spending the morning moving to a new condo in San Jose with his very pregnant wife, Jennifer, and their two-and-a-half-year-old son, Wyatt, Moss had flopped down on the sofa to watch an episode of *Law and Order* while they went out. Exhausted, he drifted off, and when he awoke from his afternoon nap, he got up to go to the bathroom. He fell flat on his face. He tried to stand up but couldn't. His legs and hands felt numb. He couldn't make them do what he wanted them to do. He called Jennifer, but she was driving through the mountains and had no cell service. So he called his mother in Port Elgin, Ontario; she wasted no time telling him to dial 911 and then headed to the airport to get on a plane to California. Moss did as he was told, and as he lay helpless on the floor, he considered his situation.

As we've seen, athletes respond to physical pain and discomfort more analytically and objectively than most of the rest of us do. Jim Moss was no exception. Instead of freaking out, he calmly took stock. Was he in agony? No. His extremities just felt like they were asleep. He ruled out a stroke, since both sides of his body were involved. Had he perhaps taken a hit on the field—been concussed and not known it—and this was just some weird delayed reaction? Maybe. Concussion symptoms can, after all, be weird and delayed. He wasn't panicked, just curious. What the heck was going on here?

Moss is a logical guy, the kind who's cautious when it comes to weighing risk and reward. He'd quit hockey after a bad concussion, because he knew another one could end his career. Rather than take that chance he'd switched to lacrosse, which at that level offered about the same pay but was safer. Like a lot of hockey players from small-town Canada, including Wayne Gretzky and Joe Nieuwendyk, Moss grew up playing both sports and had competed in the lacrosse world championships when he was nineteen. Returning to the game, then, wasn't such a startling shift, and there were some distinct benefits, including fewer games per season (sixteen, compared to eighty-eight in hockey), so he'd have time to build a career outside of sports. He wanted to have something to fall back on anyway, so when he wasn't playing lacrosse or training, Moss started selling lacrosse equipment for a national manufacturer.

But he wasn't particularly worried about his athletic career ending. A cerebral player rather than a bruiser, he had an impressive ability to think about the strategy and flow of a game and a natural talent for leadership. In both sports, he was captain or assistant captain of his teams—which is to say, he was the kind of guy coaches want to keep around as long as possible because he helped bring out the best in his teammates.

Fast forward to 2009 and Moss on the floor. This didn't feel like "good" training pain, nor did it feel like "bad" acute injury pain. Huh. He didn't start to feel alarmed until the firefighters showed up and broke down the front door of his new home. They had to. He couldn't answer it.

Doctors at Stanford Health Care swiftly ruled out concussion or any other sports-related injury. Their diagnosis? Guillain–Barré syndrome, likely the after-effect of either West Nile virus or swine flu, both of which Moss had been unlucky enough to contract in previous months. Guillain–Barré takes many forms, but in general the connection between the central nervous system—the brain and spinal cord—and the rest of the body breaks down. The good news was that the doctors had figured out what was wrong. The bad news was that, untreated, Guillain–Barré can spread rapidly and with devastating consequences: the body shuts down, beginning with the extremities but inexorably moving on to the internal organs. In the worst-case scenario, a patient would be left with a perfectly functioning mind locked inside an immobilized body. It didn't take Moss—or Jennifer, who'd joined him at the hospital shortly after he arrived—long to weigh the options. They agreed to the treatment the doctors proposed.

The plan was to use powerful drugs to wipe clean his immune system, then introduce another set of drugs to restart it: a kind of reboot for the body. The wild card? Doctors didn't know how long Moss's particular system would take to reboot. In the interim, the illness would be progressing, and if it got as far as his internal organs, Moss would wind up completely physically incapacitated; bypass machines would be required to run his vital organs. At that stage, people can die from complications. So it was a race against time to halt the progress of the illness. Once that was accomplished, Moss's condition would plateau for about a week, then (knock on wood) he'd begin to improve. In six months to a year, he'd be back on his feet. It sounded pretty bad, but Moss remembers lying in his hospital bed and feeling relatively calm for the first few days. After all, he wasn't in just *any* hospital. Stanford is a top-notch research facility (and bills accordingly—Moss jokes that they swiped his American Express card on the way in). He knew he was getting the best possible care. He had eleven MRIs in the first week alone, as well as a battery of other tests.

By day four, though, he was beginning to feel nervous. Each day, he was losing another three or four inches of feeling in his arms and legs. "I was thinking, 'Okay, I've got fourteen days left before it's into my organs.'" The idea of being trapped in a useless body was terrifying. Forget what it would do to his career—he might never be able to hold Wyatt or the new baby, still unborn, in his arms.

Fortunately, the drugs *did* take hold and his condition stabilized, then improved at about the rate he'd been told to expect. After a couple of weeks, he was ready to begin rehab. The illness plays a kind of trick on the body, forcing patients to relearn how to walk; he'd been warned that could take up to a year. But Moss didn't have that kind of time. Jennifer was due to give birth in about two months, and he was damned if he was going to be lying around in a hospital bed when she did. Yet despite his bravado, he had his doubts too. Enter the pessimistic nurse telling him he should just accept his lot, thereby triggering even more doubt. Her remark "felt really ugly," Moss recalled, "but it felt real at the same time too. And it's coming from a source that you're supposed to respect." Luckily, the second nurse arrived several hours later and said the right words to trigger his athletic instincts.

He had never been the best athlete on any of his teams, Moss acknowledged, but he could *see* the game. "I didn't have the best hands, but I could see the play making." That, plus his willingness to work a little harder than guys who had more natural talent, had yielded results. When the short, jolly nurse left, he realized he had to approach recovery in the same way he'd always approached sports. "Two things occurred to me. [First], this is the frame of mind I need. I need to feel this energized, because if I'm going to play a part in my recovery, I need a good frame of mind. And second, I can't let the next person who walks through my door, the lottery of what nurse is on what shift, change my outcome."

But how could he hang on to a positive mindset? He wanted to "hack happiness" and figure out how to insulate his current frame of mind, so he went to Google, looking for answers. The trick to maintaining a positive attitude, according to the Internet, was surprisingly simple: be grateful.

Gratitude, according to social scientists, is fundamentally important to a sense of well-being and a fully developed social life. In fact, practising gratitude—focusing on what you have to be grateful for, and making a point of expressing your thanks—has had so much success in improving people's state of mind that some researchers believe so-called gratitude interventions should be used more widely in clinical settings, to help people who are depressed or are at risk of becoming depressed.

Longitudinal studies tracking people over time have shown that when we feel and express gratitude, we also wind up accepting support from others more easily, and therefore feel less stressed. This may be why grateful people seem to cope better with change. As a group of British researchers put it in 2008, "gratitude naturally leads to improved social support and well-being during a life transition." And feeling and showing appreciation for others encourages them to provide even more support, creating a virtuous circle.

Jim Moss was ready to give it a try, especially since gratitude is so easy to feel: all you have to do is pay attention. So as he lay in his scratchy hospital sheets, eating bland food, with a cranky old man for a roommate, he began to take note of what he had to be grateful for—and he made a point of expressing his appreciation. He sent a note down to the kitchen with his food tray, saying how delicious the meatloaf had been. When the nurses came to change his rough sheets—and fresh rough sheets are more comfortable than three-day-old ones—he thanked them in a way that wasn't merely perfunctory. Who cared if he came off like a cornball? "I have nothing to lose here," he thought. "If it's corny or awkward, that's outweighed by the possible upside for me."

Almost immediately, he discovered that the British researchers were right: sincere expressions of gratitude create even more reasons to feel grateful. Suddenly a second slice of meatloaf appeared at dinner, and the nurses were changing his sheets every day, not every three days. When he buzzed for help, someone showed up right away. "I could see people kind of leaning into the positivity," he says. Meanwhile, his grouchy old roommate would press his buzzer for long stretches of time with no results, which of course only further soured his mood and made him even more unpleasant when a nurse finally did arrive. Right there in his hospital room, Moss could see the difference between the virtuous circle created by a positive attitude and the downward spiral created by a negative one.

That virtuous circle paid off in life-altering ways when he started physical therapy. Every patient was allotted a limited amount of time with the physio experts; resources were finite, even in one of the finest hospitals in the world. But Moss discovered that his determination and upbeat approach, combined with the profound appreciation he expressed for the help he was getting, now came back to him in an unexpected way. His insurance entitled him to an hour a day with an occupational therapist and another hour each day with a physical therapist. Soon, however, the therapists "started to come back at the end of their shift and donate an extra hour of their time to me," he marvels. "Then they started to come in on weekends and their days off, to give me [another] hour of their time." What was supposed to be five hours of physiotherapy a week became ten—"compound interest on positivity!" he says.

He also found that a more optimistic outlook helped him reframe his pain. He decided to view it not as a bad thing in itself, but as a form of motivation. Like Ray Zahab, he began to create thresholds for himself: how many more steps could he take today, despite the fact that each one hurt? This wasn't about martyrdom or masochism. As long as he was getting results, he could detect benefit in his own discomfort. "Optimism lets you measure more accurately the value exchange," he explains.

He wasn't just thinking about the future, though. He was also considering his own past, and what he'd learned from it. When he was thirteen, his parents were driving him home from hockey practice when a Corvette knocked their car into oncoming traffic. They were hit by a total of six cars. His mother's back was broken in two places; Moss was knocked out but sustained hardly a scratch. It was a sobering lesson in how quickly life can go off the rails, but what Moss remembers most is that everyone got better (including his mom, after six weeks in hospital), and that his family received a lot of love and support from their community. He observed resilience first-hand, and that helped him in the hospital, when he needed resilience above all else. "When you're facing a challenge and you're so deep in it that you lose perspective, if you can empathize with a former self who successfully made it through a challenge, then you can channel that previous experience into the now. But you've got to get out of the now to time-travel and see when you've gotten through hard things before."

The car accident wasn't the only hard thing he'd survived. He'd also struggled, on and off, with what he calls "mild depressions." Growing up, he was always late or losing homework assignments or being yelled at in class because he was bouncing around. Adults blamed his absent-mindedness on his obsession with sports, but Moss's take-away was that he was a bad kid, and bad at school. Only much later, as an adult, was he diagnosed with ADHD.

The challenge he was facing now, though, was quite a bit more difficult than any he'd faced in the past. The fact was that Moss was in an incredibly tough spot—still confined to the hospital, still wondering if he would fully recover, and still coping with the fact that any recovery almost certainly wouldn't restore him to pro-athlete norms. Although he'd vaguely planned on playing lacrosse for another ten years, that just wasn't an option anymore. His career was history.

There was a "new normal" to contend with, and it included significant financial stresses. Moss and his family lived in one of the most expensive zip codes in North America, and his disability insurance would bring in only about $30,000. Meanwhile, his medical bills had crept above $85,000. Not surprisingly, feelings of mild depression and anxiety set in. However, his practice of positivity made it easier to cope with those negative feelings—and to view his situation as a temporary rut, rather than a bottomless pit of despair. He thought of it this way: "A bad day is a bad day—doesn't need to be a bad week, because you can bounce back." You can also hold off on labelling any day a bad day, and try to make it better, to create what Moss likes to call stronger thresholds. "It's resilience," he says, "and it turns into a kind of micro resilience."

As the date for Jennifer's planned caesarean neared, doctors at the hospital began to joke about giving the couple a double room. "I was having none of it," Moss says. "I wanted to be out of the hospital so I could walk back into it for the birth and not be a drain on the family." His family was already feeling the strain. Jennifer—stressed out, pregnant, and caring for a toddler—had switched into survival mode, turning her attention to the home front. Moss wanted her with him in the hospital, but he felt guilty about that and couldn't find a way to tell her.

He had to get home. His occupational therapist told him that he wouldn't be ready until he could prepare a meal, by which she meant getting out of bed, showering, getting to the kitchen, and then making something to eat. He needed to prove he could manage on his own. In Moss's words, he "famously failed" the test. Although he could move, he moved slowly and painfully, and he found that in a kitchen, you're moving all the time. "You forget the mayo, you have to go back for the butter . . . we are really inefficient." For someone in his condition, inefficiency wasn't just time-consuming—it was physically excruciating. To address that, he learned tricks, like making a list of what he needed so he wouldn't forget anything, then heading over to the fridge with a bag so he could carry everything at once. He had to ration energy and be sure he made only one trip so he didn't "run out of gas."

Once he'd mastered the logistics of the kitchen, the next hurdle was the grocery store. His occupational therapist dropped him off so that he could purchase the food he needed to cook a meal. He remembers standing in the doorway, struck by how vast the store was. "Oh, my God," he thought. "This is a marathon." The key was to use the same strategy that had worked in the kitchen: figure out exactly what you need from a section of the store, then go there only once. He spent the first ten minutes figuring out where the items on his list were. Even so, walking slowly with two canes, he took an hour to complete his shopping. But he'd passed the test.

Which is how, a short time later, it came to be that Jim Moss was discharged from the hospital just in time to walk back in for his daughter's birth on November 24, 2009.

We may be wired as a species to avoid discomfort, but athletes like Ray Zahab and Jim Moss know to embrace it. They seek it out because they intuitively understand that if they challenge their bodies, their bodies will respond not by collapsing but by adapting and growing stronger. In other words, the right degree of stress will build resilience. In fact, the human body adapts so well to hardship that sometimes it flourishes more under trying conditions than when it's coddled. Studies have shown that when lab animals are underfed or exposed to toxins, limited radiation, or uncomfortably high temperatures, they often live longer than those cossetted in stress-free environments. It's as if a certain level of stress inoculates them—what doesn't kill them really does make them stronger.

There's a name for the tendency of living organisms to thrive when the going gets tough: hormesis. Jim Moss may not have known what it was called, but he was certainly well acquainted with the phenomenon. If he weren't, he'd never have managed to become an elite athlete. It was because he possessed an athlete's mindset and had struggled to achieve excellence in sports that he knew to push the envelope when his mobility vanished. Athletes who don't push themselves lose their edge, and once that happens, their performance suffers. And then it's game over. But Jim Moss didn't want his performance to suffer. Jim Moss had goals.

Sometimes those goals were too ambitious. On one warm and sunny California day in January 2010, Moss decided he would try to go for a run. He'd relearned to walk, after all, and was walking laps around a local park. But the moment he tried to accelerate to a jog, he lost his ability to stabilize himself. He fell, again and again. It was a heart-breaking experience for a lifelong athlete. His occupational therapist had warned him about pushing too hard, telling him that at some point, more effort would not yield more results and could actually set him back. There was a limit, he was finding, to his ability to accelerate his recovery. Nature was taking over—at its own speed.

But that didn't really put a dent in his willingness to push himself as hard as he could. A week or two later, he went to a San Jose Sharks game. He assumed that because he could walk laps around the field and would have crutches for help, he'd manage just fine. He could get around the grocery store easily now, so how hard could an arena be? The answer: unbelievably hard. He'd forgotten that ten thousand excited (and in some cases, inebriated) people would also be there. "They're just knocking you all over the place," he says ruefully. And there were emotional knocks too. "This is the actual building that I played hockey in. Going from being the number-one player on the floor to being the guy who can't even manage to get to the mezzanine level . . . that hit hard." It was a sharp, sour reminder that as far as he [illegible] come, he still wasn't where he wanted to be. And he'd pro[illegible] ably never get back to where he was.

Subsequently, Moss fell into a funk and had "a few dark weeks." To pull himself back up, he turned again to the strategy that had worked so well in the hospital: reframing his unhappiness as a prompt to remember to count his blessings. "I went back to gratitude." The first thing he felt he had to be grateful for was having the time to get better. And he wouldn't have had the time if the treatment hadn't worked, and if he hadn't worked so hard at rehab. He could look at his athletic prowess as something valuable he'd lost, or he could look at it as wonderful preparation for the biggest test of his life: full recovery. There was a lot to be grateful for, actually. His spirits lifted.

Once he was back on his feet, both literally and metaphorically, Moss was amazed to discover that his homespun solution was validated by research. Looking on the bright side worked not just for him but for countless others as well.

Once upon a time the assumption was that if you wanted to be happy, you had to have something wonderful happen to you. Maybe you'd fall in love. Maybe you'd score a big inheritance. It didn't really matter what brought you happiness. The point was that there had to be a precipitating event for it to rain down upon you. But in the late sixties, University of Pennsylvania psychologist Martin Seligman started to wonder whether the cart hadn't been put before the horse. What if, in fact, happiness was brought on not by our circumstances but by our attitude? It was an idea that ran counter to the prevailing notion of behaviourism in psychology, but Seligman ran with it, developed a groundbreaking theory around it, and backed it up with science. The positive psychology movement was born.

Seligman believes happiness doesn't have all that much to do with external events in our lives. If it did, lottery winners should be a whole lot happier than car crash victims who've been rendered paraplegic. But they aren't, according to an oft-cited 1978 study, which found that on the whole, one group wasn't any happier than the other. Happiness, Seligman argues, is an individual choice. We choose to cultivate an attitude that promotes it, or we don't.

You know those people who always seem to be relentlessly upbeat and cheerful? Seligman studied them and found that most share certain signature strengths—but he also found that anyone, even the worst curmudgeon, can cultivate those traits. Not that turning yourself into a more positive and optimistic person is a cakewalk. You have to do the work, Seligman warned. But if you do, you'll be rewarded: in all likelihood, your happiness level will increase. (So may your longevity, apparently. One study that looked at two groups of nuns who led virtually identical lifestyles found that those who expressed positive emotions more frequently and intensely in their daily journals outlived many of those who did not.)

To achieve greater happiness in the present, Seligman found, you have to think constructively about the past and optimistically about the future. The key to both exercises is learning to feel gratitude and forgiveness. Gratitude and forgiveness are, in other words, the twin pillars of happiness. They are also misery's one-two punch.

To some extent, our ability to feel grateful may be genetically predetermined: scientists have identified a gene variant associated with gratitude. But nurture definitely plays the largest role—you can learn how to feel more gratitude. Research suggests that it's a bit like learning how to play the piano: if you practise feeling grateful in a disciplined way—let's say by making a regular habit of recording all the reasons you have to be grateful, then spreading it around a little by expressing your appreciation to the people in your life to whom you feel grateful—you'll feel happier than grousers and complainers do.

As Jim Moss discovered, and his grumpy hospital roommate unfortunately did not, choosing to focus on the reasons you have to be grateful instead of the reasons you don't not only improves your happiness quotient but also brings out the best in those around you. A 2012 study by researchers at the University of Southern California found that a simple thank-you goes a long way toward placating a testy or unreasonable boss. When a powerful individual's competence was questioned or challenged in any way, that person tended to lash out aggressively; gratitude, however, disarmed and calmed him.

So why was Moss so good at cultivating gratitude, but his hospital roommate was not? Seligman wrote in *Learned Optimism: How to Change Your Mind and Your Life*, his 1991 classic, that after studying optimists and pessimists for twenty-five years, he'd concluded that pessimists tended to believe bad events would last a long time and would undermine everything they did. Optimists, on the other hand, when "confronted with the same hard knocks of this world," thought about their misfortune in a radically different way. They tended to view any defeat as a temporary setback, and they remained unfazed by it. And unlike pessimists, who blamed themselves when things went wrong, optimists believed that "circumstances, bad luck, or other people" had brought about hardship, so they didn't spiral into negativity. Rather, they reframed the discomfort of defeat as a temporary challenge—and tried harder to overcome it. Just as Ray Zahab and Jim Moss did.

In the spring of 2010, when their daughter was just a few months old, the Moss family returned to Waterloo, Ontario, for a short visit with relatives. They never left. One thing that kept them was just how grateful they felt to be there. "There was an extended theme of gratitude—for the health care system, [for] having our parents close by, for the cost of living," Moss remembers. Jennifer, who'd been a public relations executive at a major employee placement agency in California, quickly found a new job in Waterloo, and they moved into a new home. Meanwhile, Moss continued his slow but steady recovery.

A self-confessed energy junkie, he'd been calibrated for non-stop action. But now, it took him forty minutes just to walk around the block. Rather than allowing himself to feel frustrated, he focused on "all the good things about going slow." Time to smell the roses, time to talk to his kids, time to meet a neighbour or two. "The difference between optimism and pessimism," he explains, "and why gratitude is so integral, is that when you're grateful for things, you're actually making lists of the resources that you have. *All the time.* Pessimism has a lot to do with 'what can't I do because of what I don't have?' And optimism is [about] 'what can we do with what we've got?' Gratitude is about making a list of all the things you have." It's harder to feel unhappy when you're aware of how much you have and you feel grateful for it.

The sheer amount of time he had on his hands now that he wasn't working might have depressed Moss. But instead, he felt grateful for it. He could be present in his young children's life, and he could also do things he had long wanted to do, like building a computer from scratch and reading books. After a while, he realized that what he really wanted to do was fill his time by going back to school to study happiness and gratitude.

Though he'd never loved school—sitting still was challenging for Moss, so much so that he'd turned down a hockey scholarship from Cornell University years earlier—he was hungry to know more about how happiness works. "I had a really interesting thing to study," he says with his broad smile. "Basically myself." But the real question he wanted to answer took him back to that hospital room at Stanford: "Why me in one bed, feeling gratitude, and that grouchy old guy in the next? Why do some people get cancer and they break, while others beat it and go on to do incredible things? What part of that is innate, what part is the environment, and what part is entirely out of our control?"

In September 2010, Moss enrolled at Wilfrid Laurier University to study psychology. While there, he came across the concept of post-traumatic growth (PTG). The inverse of post-traumatic stress, PTG is about turning adversity into positive psychological and emotional change. Researchers were finding that six months to a year after a life-changing traumatic event, people with PTG wouldn't change that event even if they could. They'd found value in the event in the form of growth. Moss related to that because the same thing had happened to him, and emotional growth had cushioned him from the impact of some serious blows. For instance, one of the biggest and most shocking changes in his life had been the abrupt end of his athletic career. But rather than mourn the loss, he'd been philosophical, reframing it as an opportunity to move in a new direction. "I had already made the all-star teams, won a world championship . . . so the only real next milestones had to do with longevity [as an athlete]. I wasn't going to go back and do those 'firsts' again. I was able to see that, partly as a way of helping myself understand that it was okay to put that chapter to rest."

He'd grown in other ways too. In school again at age thirty-three, the self-described "creepy old guy in the front row" was a completely different student than he'd once been because he understood and was able to manage his ADHD now. He knew what his learning challenges were and, as he puts it, "hacked them," standing up if he needed to move, for instance, and studying in busy places that kept his brain active rather than morgue-like settings.

As he continued his studies, he did a deeper dive on the significance of gratitude, via a blog he created that invited others to share pictures of things they were grateful for. He had thought deeply about why *The Smile Epidemic*, as he called it, might help people feel better. Interacting via technology with other people made gratitude a physical, concrete act, and by sharing their reasons for gratitude with others, participants found they reinforced their own appreciation.

Moss was seeing the same thing at home. At the end of each day, he and Jennifer had got into the habit of asking Wyatt what had made him smile that day. The first couple of times he was asked, the toddler looked around and named a person or an object that happened to be within sight. But within just a few days, he began to name things that had happened much earlier. In other words, because he knew the question was coming, he was noting things that made him happy and creating his own gratitude list. And since our brains have only so much capacity for memory and attention, this almost certainly meant that Wyatt was starting to look past or ignore the not-so-positive events of the day.

Do people who are grateful wind up having better lives, or are they grateful *because* their lives are already so good? When researchers Robert Emmons and Michael McCullough set out to try to answer this chicken-or-egg question, their hunch was that gratitude caused happiness. They hypothesized that active gratitude—that is, doing the kinds of things Jim Moss was doing, like making lists of what he felt grateful for—induces a heightened sense of well-being, compared to focusing on the negative or even thinking about neutral events.

To test their theory, Emmons and McCullough split 192 university students into three separate groups and asked them to record their moods, behaviour, health, and life satisfaction at the end of each week for ten weeks. The students in the first group were given a statement suggesting that there are many things in our lives to be grateful for, big and small, and they were asked to think back over the past week and list five things they were thankful for. Their answers ranged from "the generosity of friends" to "the Rolling Stones." The students in the second group were given a prompt suggesting that there are a lot of hassles in life, and they were asked to think back over the past day and list up to five hassles; their lists included items such as "stupid people driving" and "a horrible test." Those in the third group were simply asked to think back on the past week and write down events "that had an impact on [them]," and their answers included "learned CPR" and "cleaned out my shoe closet."

The students were also asked to rate their moods and any physical complaints (headaches, stomach aches, and so forth) at the end of every week, and to summarize how much exercise they'd had and the level of social support they'd received. (This last question was important, because gratitude is generally a response to positive interaction with another person.) Finally, they had to rate their overall well-being and briefly outline their expectations for the coming week.

It quickly became very clear that those in the gratitude group were faring better, and felt better about their lives, than those in the other two groups. They were also "more optimistic regarding their expectations for the upcoming week," and they "reported fewer physical complaints and [spent] significantly more time exercising"—an hour and a half more each week than those in the "hassles" group.

But the researchers wondered if their results had been skewed by their sample. After all, most university students are relatively privileged; those in the study were young and attending top American institutions. Presumably, their reasons to feel grateful were plentiful. So Emmons and McCullough decided to ask sixty-five people between the ages of twenty-two and seventy-seven who had neuromuscular diseases to spend about five minutes each day for three weeks rating the same kinds of things the students had been asked to rate: physical complaints, overall well-being, how much sleep they were getting, and so on. Half of the new subjects were also asked to list a few things they'd felt grateful for that day. At the end of the study, the participant's spouse or significant other was also asked to evaluate his or her partner's satisfaction with life.

Once again, the participants who'd been cued to think about gratitude each day "reported considerably more satisfaction with their lives as a whole, felt more optimism about the upcoming week, and felt more connected with others" than did those in the control group, who were simply reporting on their day. What's more, those who were logging what they felt thankful for were also getting more sleep and woke up feeling more rested than those in the other group. The researchers concluded that simply reflecting on gratitude "led to substantial and consistent improvements in people's assessments of the[ir] global well-being"—and these improvements were so noticeable that the spouses of those in the "gratitude" group rated their partners as more satisfied with their lives than did the spouses of those in the control group.

Counting your blessings, apparently, really does work. Even if you don't have all that many of them, you will feel a lot happier if you spend a few minutes every day actively figuring out what they are. Moss stubbornly insists that everyone does have great moments, even on very bad days. "When I was in the hospital—and couldn't walk, and couldn't pay my mortgage, and couldn't help Jenny, who was about to have a baby—banana pudding made it great today." Even when everything feels bad, in other words, there's probably at least one good thing to celebrate. You may just have to look harder for it.

Moss's wakeup call about his own future came when he realized how much research there was on the subject of gratitude, happiness, and retraining our brains to experience more of both, yet how little of that research had been popularized. Most people didn't even know it existed. The vast majority of it was sitting on library shelves, in the form of doctoral theses, gathering dust. Moss, who had been planning to get a PhD, realized, "That's where mine would end up too." And so he made a decision: he would forego the PhD and try to monetize the knowledge that already existed. He'd build a business based on gratitude.

Plasticity Labs was launched in 2013 with a clear mission: "Helping millions of people find the tools to live a happier life," says Moss. He didn't want to charge individuals for those tools, though. Instead, he thought he'd build apps for corporations, which would "pay us to do what's good for their people—and also good for them."

Moss's first client was a Waterloo-based company called Unitron, which sells hearing devices. Sales were suffering, and Unitron's CEO thought customer service, then dismal, could be drastically improved if only staff members were happier. Moss and his team of five PhDs—and Jennifer, who had quit her job to help launch the business and is now the chief marketing officer—recognized that the big opportunity wasn't in creating software customized for individual businesses, but in creating one tool that could work for many companies.

By 2016, Plasticity Labs (PL) had dozens of customers ranging from a bank to a branch of the federal government, and revenues were growing at 400 percent a year. PL supplies a workplace social-networking app designed to foster resilience and optimism, and also performs surveys to gauge employees' happiness. In addition, Moss and his team offer group and individual training and activities tailored to a company's needs. For one recent client, for instance, PL helped employees prepare for a major growth spurt.

The app works like this: employees voluntarily log onto the Facebook-like portal on their mobile phones, rate their happiness for that day on a scale of 1 to 100, and provide a reason for the score. Anonymity is guaranteed; all the data PL receives from employees is "washed" so that each user's unique identifier is replaced with a universal one. (Not even Jim Moss knows who said what.) Employees can also use the app to network online to share successes (and feelings of gratitude). This social-networking aspect seems to keep people coming back, and the app also offers surveys to measure traits that are seen as precursors to happiness, like resilience.

Wording is key, as Moss learned from that nurse back in Stanford, so users of the app are asked, "What made work great today?" rather than "How was work today?" There's an explicit presumption that something did make work great, which is a way of "priming the brain" to find those moments.

Some employees—25 percent, according to a PL survey—even choose to engage with the app on their own time. Moss believes this happens because the skills and traits the app promotes—empathy, resilience, optimism—also pay dividends at home. Most people, he contends, want to grow and change and become happier, but many are self-conscious, and for them, an anonymous survey or training course is a godsend. "If I ran a class on optimism in the office, the people who would come are the people who don't need it," he points out. "But if I give you a private place to do it and no one knows you're there, it's incredible how many people lean into it."

By giving employees a scale against which to measure their own happiness, the app also offers them an opportunity to gain some perspective on their own lives. Let's say one day an employee had a lousy commute and logged a 60 out of 100 on the happiness scale. The next day, that same employee learned that a friend had been diagnosed with cancer. What number would he record then? Would he rethink the previous day's 60? Almost certainly. When people have a longer-term perspective on their own happiness, they also have a more accurate barometer of discomfort and their reaction to it. Maybe that long commute really wasn't so painful after all; maybe it could be reframed as an opportunity to prepare for the day's meetings—or just an opportunity to daydream.

Gratitude doesn't figure prominently in the culture of most companies, but it should, Moss says, because focusing on the things you're grateful for at work has a significant positive effect on job satisfaction. In one study conducted by Plasticity Labs, sixty-five employees at a variety of companies, ranging in age from eighteen to eighty-two, engaged in "gratitude activities": half of them described three things they were grateful for at work, while the other half described three things they were grateful for in life. All sixty-five also filled out surveys about how satisfied they were with their jobs, how they thought they'd feel about their work in six months, and how much sense of community they felt in the workplace. Those employees who'd focused on three things they were thankful for at work (among other things, they mentioned gratitude for their co-workers, job flexibility, and health benefits) reported more workplace gratitude than those who wrote more generally about feeling grateful in their lives. They also anticipated feeling more job satisfaction six months down the road—a different version of the virtuous circle Moss himself experienced when he focused on gratitude in the hospital.

Just as fear spreads like wildfire once rumours of an organization's financial troubles start circulating, a culture of gratitude can go viral too, with a big impact on morale and productivity. "Grateful people communicate better, handle the challenges of life better, [and] have more positive relationships, more positive levels of perceived social support, higher self-esteem, and a [greater] sense of self-worth," Moss explains.

The second important element of Plasticity Labs' business model is that it offers companies a treasure trove of anonymous real-time data on their employees' states of mind. A business can't get its hands on an individual employee's responses, but by tracking trends, companies can adapt and provide relief when necessary. If last year's data showed that late August through early September was a time of high stress for many employees, a company might adjust its hours, for instance. Julie Dopko, now director of human resources at Unitron, Plasticity's first customer, says that getting employee feedback so quickly, rather than waiting months while data from an annual survey is crunched, is a huge advantage because the company can respond without delay. Her previous employer, which was Plasticity's second customer, discovered from a PL survey that employees wanted more time together as a team—they just hadn't felt free to say so to senior management. So the company promptly instituted weekly team lunches, organized monthly outings, and emphasized group work wherever possible.

For Moss, mindfulness is the key element he's trying to introduce into people's lives. Meditation is one way to do that, but he believes simpler methods can be just as effective. For instance, he says, "Stop and ask yourself, 'What do my pants feel like against my legs right now?'" It's a short step from that to thinking about how your pulse races or your temperature rises when you're about to lose your temper. Really tuning into your feelings and surroundings a few times a day can instill a habit of awareness, of being in the present. "Part of people's initial disbelief about this is that they don't think it can be so simple yet have such a profound effect," he explains. "But that's the nature of happiness: one simple change in how you think about happiness can have huge impact."

Plasticity Labs is currently working with elementary schools on a pilot project, and Moss dreams of a whole generation of young people who, by the time they reach the workforce, already have the ammunition they need to reframe discomfort: gratitude, mindfulness, and happiness skills. "Basically, I'd like to put us out of business," he says with a laugh.

CHAPTER FOUR

Rejecting Discomfort

When you want to—or have to—change, you can deal with the discomfort of it by going the Ray Zahab route: reframe it as motivation and convert it into the fuel you need to power through. This is what athletes learn to do to strive for better results. Discomfort can be a tool we use to help us change for the better or overcome an obstacle. But what if the mind you're trying to change isn't your own? Let's say people don't think much of you, or are actually rooting for you to fail. How do you win them over? Discomfort that comes as an external force emanating from other people can be one of the most difficult types to deal with.

The answer is tied to depersonalizing criticism, even—especially—when the criticism is very personal. This reframing technique is the opposite of mindfulness: instead of drilling down deep into the discomfort, you deny that it even exists and don't allow yourself to experience it on an emotional level. You do, however, take it on board intellectually, using it as a challenge to prove your worth and overcome the bias—thereby silencing your critics.

It's a strategy that seems to have worked quite well for Linda Hasenfratz, a pioneer in a male-dominated field, and someone who got there—at least at first—because her father owned the company. Taking over the family business might sound like an easy path to success, but this was a sizeable auto parts company. On a factory floor in 1990, with its huge machines noisily hammering out high-precision metal parts, you didn't see many women working the lines—or doing much of anything else, for that matter. So blond-haired, blue-eyed Linda Hasenfratz definitely stuck out. It didn't help that she was twenty-four years old and had no training relevant to the industry; her degree was in chemistry, and her first job had been selling pharmaceuticals. But she'd decided she'd rather take the chance to run her dad's company—a big leap not only because she knew next to nothing about auto parts, but also because the corporate culture was rooted in performance. She would need to prove she could perform. Employees had a right to be skeptical. Linda Hasenfratz didn't look like anyone's idea of a factory boss.

She was, however, the daughter of a man who'd taught her a thing or two about determination. Frank Hasenfratz was a Hungarian refugee who fled to Canada after the Hungarian Revolution in 1956 and eventually started a business in the basement of his Guelph, Ontario, bungalow. He called it Linamar—for his daughters, *Li*nda and *Na*ncy, and his wife, *Mar*garet—and set out to supply parts to the aerospace and auto industries. By the time Linda joined in

1990, Linamar had nearly nineteen hundred employees and seven plants in Southern Ontario; though each plant functioned as an independent entity, the overall company was a well-oiled machine that reflected the single-minded drive and sense of purpose of its hard-working founder.

Although she'd grown up literally next door to the original Linamar factory in Guelph, Linda Hasenfratz initially had no intention of joining the family business. She had no particular passion for auto parts, and she wanted to see what she could accomplish on her own. In high school she'd excelled in science and math, and in university she'd studied chemistry—graduating at the top of her class, despite never missing a party. But after that stint as a sales rep for a pharmaceutical company, she reconsidered the appeal of auto parts. "I realized Dad was running a business he'd built from scratch into one with a hundred million dollars in annual sales. I thought, 'If I play my cards right, I have the chance to run a hundred-million-dollar company one day.'"

Frank Hasenfratz remembers the day Linda called him and his wife on vacation to say she had something to tell them when they got home. They spent the next two days conjuring up worst-case scenarios. What could be so awful that she couldn't bear to tell them on the phone? It never occurred to either of them that she wanted to join Linamar, with the goal of running it one day. When Frank got home and heard his daughter's big news, he was surprised. He had neither expected nor encouraged his daughters to work in the business he'd built. In fact, he says, when he took Linamar public in 1986, he did so thinking he'd turn the reins over to an outsider when he was ready to retire.

So he was delighted by Linda's announcement, but also cautious: this was his life's work, and thousands of employees who'd helped him build Linamar were depending on him. Family was everything to him, but he couldn't hand Linda the keys just because she was his daughter. Instead, he wrote out an eight-year plan that would give her first-hand exposure to every aspect of the company: she'd work in just about every job short of machining to get ready to run the place someday—if she proved she was up to the task. "She would know herself by the end if she could do the job," is how Hasenfratz senior puts it.

Thus it was that Linda Hasenfratz found herself moving from the shop floor to quality management to materials management, and so on, while rotating through Linamar's various plants. In some companies, the family connection might have invited deference or at least a modicum of politeness. But that wasn't always Hasenfratz's experience. Though she is personable and friendly, she encountered some thinly veiled hostility from co-workers who understandably believed nepotism was the only reason she'd been hired. Others treated her as though her gender rendered her irrelevant, even invisible. From the start, it was clear that no one on the factory floor was going to make it easy for her.

Nor would time help much. No sooner had she made inroads in one plant—having learned, say, how to make steering knuckles, the part of the wheel that attaches to a car's suspension—than it was time to start all over again in another plant and learn how to make airplane parts. Each time she moved, she encountered a new group of skeptical and judgmental fellow employees, some of whom were probably hoping she'd fall flat on her face. Hasenfratz tried not to let that rattle her, taking the approach that she was bound to make mistakes. And she did. "The trick is [not to] make the same mistake twice," she says today. "Learn the basics first. Work on perfection once you have the big pieces in place."

She wasn't demoralized by the occasionally frosty reception and she didn't take it personally, though it was personal. Instead, she reframed instances of sexism as legitimate concerns about quality control and redoubled her efforts to do her best work; she repackaged whispers about nepotism as reminders that she had a special responsibility to uphold her father's performance standards. From the start, she'd made a conscious choice to turn down the dial on her co-workers' judgments and personal slights, the way she did on her car radio when she got bad reception. She'd listen to the music, not the static.

It's the same advice she gives to young women today: "If you go looking for a negative opinion, you're going to find it. Just ignore it. *Don't look for it.*" And when it comes anyway, don't take it personally.

Tuning out personal criticism came easily to Hasenfratz, possibly because she was raised by a father who cared more about performance than personality or status, and a mother who had a tendency to look for the best in people. Hasenfratz found that if she behaved as if the personal animosity in the plants simply didn't exist, focusing instead on working hard without ever asking for special treatment, people "quickly forget about all of it—that you're young, a woman, the boss's daughter—and you earn their respect." Her father approvingly says, "She doesn't expect people to adjust to her; she adjusts to them."

Repeatedly facing the same uncomfortable environment—and having to find a way to succeed within it—cemented her ability to respond to hostility with equanimity. She can laugh about it now—and does, in her infectious fashion—but at the time, her co-workers' antipathy wasn't so easy to brush off. With each new job or plant, she'd think, "I just spent six months getting the last team on board, showing that I can do this, and now I have to start over." At one point she recalls feeling "a bit disheartened," but her father gave her good advice that helped her persevere: "Just hang in there and don't get discouraged. Tough times make you tougher." There were no long discussions about the difficulties she was facing, and no coddling either. She was expected to buck up and deal with the skeptics by proving them wrong, so she did.

The skill Hasenfratz acquired—of changing the frequency when she didn't like the noise around her—was an adaptive one that has served her well throughout her career. In 1997, when she was thirty-one and managing some of Linamar's manufacturing plants, the president of the company unexpectedly stepped down. It wasn't the time they would have chosen, but she and her father agreed it made sense for her to leapfrog into the role of chief operating officer. Overnight, she became the boss of a good number of people, including several senior managers to whom she had reported literally the day before. "That was the toughest transition because it didn't follow the right flow," she says. And while she acknowledges that the situation was "a little awkward at first," her response was to "just muscle through it." She had to change everything, from what she wore to work—more corporate attire was now in order—to the way she interacted with her former bosses, who were now direct reports. Confidence and humility were required.

"You don't have to be 100 percent ready for every job before you take it. Being a little not-ready makes you stretch and grow and find your way into the role," she says. But her father was pretty sure she *was* ready. "Would I really risk everything I had built just to bring Linda in?" he asks rhetorically. "I would never have done that!" It helped, he said, that she hadn't ever acted entitled or tried to pull rank. She genuinely liked the people working the plant floor, he says, and ate lunch or dinner with them rather than retreating to the executive offices.

It was this true understanding of the company and its people that qualified her for the executive suite, in her father's eyes. But her job now was to prove herself capable to people who didn't share her last name, and to take the company to the next level. That would require changing some of the business practices her father had put in place—business practices that were, in fact, working rather well in individual plants but would need to be standardized and adapted as the business grew—and outsiders weren't at all sure she was the right person for the job. History books are crammed with children and grandchildren who steered their inherited businesses right onto the rocks. As one analyst who follows the auto sector put it, "Any time a family business transitions from first to second generation, people get nervous." It wasn't just nepotism that made outsiders worry about

Linda Hasenfratz: they weren't sure she had the business and management chops. Where was the proof? "In the early days, there were a lot of questions about her leadership," said David Tyerman, then a managing director at Canaccord Genuity. "In the 1990s, Linamar was a much smaller company and growing quickly. There were costs related to that growth which held back results, and people thought she got the job because she was Frank's daughter and that the results were terrible."

But Hasenfratz had a vision for Linamar: she wanted to bring it into the modern era, then expand significantly. To get there, she needed to persuade others that change would be good for Linamar—and here the knowledge she'd gained on the shop floor, the relationships she'd forged there, and the credibility she'd earned made all the difference. The first thing she did was to create standardized practices to make the whole operation more efficient. Because the business had grown organically, it had a strong entrepreneurial culture, with plant managers thinking and acting like owners themselves; on the downside, the plants all ran with different systems, so something as simple as comparing performance was difficult. Standardized reporting allowed Hasenfratz to push improvements in operations even further, in a manner that reinforced the performance-based culture that was so important to Linamar's corporate DNA.

Hasenfratz was also keen to keep growing the business, so she doubled down, investing in new equipment and more tech-savvy workers while also expanding into new markets, like Europe and Asia. Her timing was good: as she was working to make the disparate plants of the old Linamar function in a more modern way, Linamar's auto-industry customers were also changing and modernizing. For a long time, the biggest automakers had designed and machined most of their own parts, farming out only the low-end bits that were easy to manufacture. But in the early 2000s, they began to shift more of their production out of house, and Linamar, along with fellow Canadian auto-parts manufacturer Magna International, was there to capitalize on the trend. Hasenfratz's investment in the company paid off, enabling Linamar to corner a big share of the market for powertrains, the part of a vehicle that transmits the power for motion from the engine to the axle of the car.

And there was another shift: automakers were now willing to let parts suppliers in on the research and development of products. That gave Linamar and Magna a more lasting advantage because it allowed them to build R&D know-how and gave them deep insight into what automakers needed. Hasenfratz's vision was paying off. In 2001, Linamar was providing $60 worth of parts per vehicle, on average, in North America; by 2010, the figure was $130 worth.

Although his daughter was changing the company he'd built, Frank Hasenfratz approved of her approach. "An entrepreneur can take a business to a certain level, but if you want to go global, it often takes a different person," he told me. "She was that person. When she wanted us to go to China I said it was too early, and she said, 'I'm doing it.' Our agreement was I could comment, but the final decision was hers. Now China is our most profitable subsidiary."

But there were still those who doubted her ability. When Hasenfratz decided to diversify by acquiring the construction equipment manufacturer Skyjack in 2007, some feared that it would distract Linamar from its core business. She fields questions about the diversification play to this day—and has had plenty of opportunities to "turn down the dial" on the sniping critics—but she asserts that having alternative sources of revenue has helped the company. There's no denying that it hasn't all been smooth sailing, however. When the credit crisis and global recession hit in 2008, construction companies like Skyjack took a beating—and so, of course, did auto-related businesses. Linamar's biggest customers were on the ropes; two of the Big Three, General Motors and Chrysler, filed for bankruptcy protection and needed massive government bailouts just to survive. Combined, they owed Linamar $30 million.

Linamar also went into survival mode, cutting costs rapidly—and painfully, with salary cuts for management and massive layoffs. That wasn't easy in a close-knit company, but Hasenfratz chooses not to dwell on unpleasantness, saying only, "We had to make tough decisions. There is nothing that feels good about taking someone's job away." The company managed to pull through, and as weaker competitors faltered, Linamar even snapped up some of their business.

The downturn actually became a time of potential, not disaster. Though still dwarfed by Magna, which had sales of $32 billion in 2015 and plants in twenty-five countries, Linamar has grown substantially, from under $1 billion in sales in 2002, when Hasenfratz took over as CEO, to $5.2 billion in 2015—a 24 percent increase from the previous year. That's still not big enough to suit Linda Hasenfratz, though, who says she's on track to reach her goal of $10 billion in annual sales by 2024.

But despite her impressive track record, industry analysts and journalists alike rarely fail to mention that she's the boss's daughter. Almost twenty years later! She deals with the implication that her success is due to nepotism in the same way she copes when she's challenged on strategic decisions inside the company: she just tunes it out. I saw traces of this myself when I asked her about sexism and the difficulties she faced in her career. She couldn't think of any specific examples. Yes, she conceded, there probably were instances of sexism. She just couldn't remember any. "Right up until today, maybe someone is questioning my ability," she says with a shrug, "but I'm just not dialling in to that."

Hasenfratz did, however, always dial in to the fact that she was the boss's daughter, and she understood that it could be a huge plus—but only if she owned the responsibility that came along with it. After all, the boss's daughter has an opportunity to learn the ropes in a way that no one else ever could. Instead of using her family connection to land a cushy position, she took advantage of the access it created to learn every corner of the business, sweating it out on the factory floor alongside the people she now leads, so that she'd be uniquely qualified to be CEO. She attended to the genuine concern underlying the gossip—*we need a strong leader at the helm of this company*—but ignored the snark and the sneers, thereby denying their power to hurt her.

By earning her stripes the hard way, she earned the respect of detractors and was able to build the alliances and relationships she needed when she embarked on a major expansion and transformation of the company. And by not taking criticism personally, she managed to turn a negative into a positive. Yes, she was the boss's kid—but if she weren't, she might not have been able to turn herself into a visionary leader.

People often think that if you're in denial, you're too weak, obtuse, or delusional to face up to reality. It's easy to see why. If you have a problem you don't want to face—you're eating yourself into an early grave, say, or your relationship is destructive—denial tends to make matters worse. Rationalizing a big problem right out of existence may allow you to maintain a sense that all is well and you're in control of your life, but in the long run, it's a strategy that will almost certainly wind up causing considerably more damage. For instance, when Steve Jobs was diagnosed with a rare form of pancreatic cancer, he refused potentially life-saving surgery for nine months on the grounds that he didn't want his body to be violated in that way—a view that accorded with his Buddhist beliefs. Walter Isaacson, his biographer, reported that Jobs told him later he regretted his choice and believed he'd engaged in magical thinking. When that news became public after the Apple founder's death, many people wondered how such an intelligent person could have denied the severity of his illness and made such an apparently delusional choice. But Jobs was a master of the universe who'd often tuned out other people's advice to go his own way, and it had worked out rather well for him. So why wouldn't denial and magical thinking work when it came to illness?

Denying the discomfort of change, however, is another matter. If you're already attempting to change, you've faced up to reality and decided to take some kind of action. Denial, then, is not a refusal to act—it's another way of reframing discomfort. Rather than inspecting it non-judgmentally, as Andy Puddicombe learned to do, or parsing its meaning, as athletes like Ray Zahab are taught to do, you ignore it altogether so you don't get distracted by it. That's why Linda Hasenfratz responds to queries about the difficulties she encountered in her climb to the top with a quizzical look: they didn't stick in her mind because she chose to ignore them.

Freud argued that denial is a defence mechanism that allows us to ward off an external reality threatening our ego, or sense of self. Denial is not always bad, in other words, and can be a handy psychological tool that lets us ignore what we can't control and what would only hurt us. If Hasenfratz had let the snide comments get to her, it would have been harder to keep going from one plant to the next, learning the ropes. And if she'd listened to the naysayers, it would have been much harder to grow her business. Even if she'd succeeded, she would probably have felt a lot more insecure and unhappy along the way.

Along with the bad kind of denial, which prevents you from taking action to fix a problem, then, there's the good kind of denial, which enables you to keep plowing ahead even when a situation is extremely uncomfortable. In fact, many psychologists now believe that denial can be enormously helpful as a short-term strategy because it buys you time to process a difficult or traumatic situation. (Joan

Didion, nobody's idea of a truth-denier, titled her bestselling memoir about adjusting to the sudden loss of her husband *The Year of Magical Thinking*. Denial, clearly, helped her keep going.)

If Linda Hasenfratz had bristled at every slight, she couldn't have won over the rank and file at Linamar—and once she was in the corner office, she couldn't have led them effectively (not least because they might have been terrified that she'd seek revenge). Because she responded to them as though they weren't criticizing or insulting her, they . . . stopped. Because she treated them as though they had no problem with the fact that she was the boss's daughter, they . . . didn't.

From a mental health standpoint, denying negativity is definitely preferable to dwelling on it obsessively, which psychologists call rumination. Unlike mindfulness—where, as Puddicombe explained, you're saying, "Oh look, there's a thought," and trying to separate the content from the emotions surrounding it—rumination is grabbing hold of that thought and judging it, endlessly mulling the causes, meanings, and implications of bad feelings or negative past events. The purpose isn't to move past the event but to relive it, over and over. Rumination, as a wealth of research has proved, is bad for us: we are likely to wind up in a downward spiral, and are much more likely to develop symptoms of anxiety and especially depression—four times as likely, according to one survey of thirteen hundred people aged twenty-five to seventy-five.

Rumination appears to make bad situations even worse. A study of Bay Area residents who lived through the 1989 San Francisco earthquake found that those who ruminated experienced more symptoms of depression and post-traumatic stress disorder. Similarly, a study of 455 people whose family members had died after terminal illnesses showed that those who ruminated were more likely to become depressed, and eighteen months later, most were still depressed.

But even when ruminators are not mulling over something negative, the mere act of rumination makes them feel worse, according to a plethora of studies. When people are given an emotionally neutral statement, such as "Think about the level of motivation you feel right now," or "Think about the long-term goals you have set," then are asked to focus on their feelings about it for eight minutes, those who are prone to ruminate feel worse at the end of the exercise, while people who don't ruminate feel the same way they did when they began. By contrast, when researchers instruct people to spend some time thinking about the blades of a fan or the layout of their local mall, ruminators feel better afterward—they've been distracted—while non-ruminators feel just the same.

If ruminating on something negative feels bad, and is so bad for us, why do we do it? Well, dwelling on a problem can create the illusion that you're doing something about it. After all, look at all the time you're putting in. You must be getting *somewhere*! But ruminating is in fact a way of spinning your wheels and wasting time. The more you spin, the more stuck you get—and the deeper your rut becomes.

Dwelling on a problem seems to impair our ability to solve it in some fundamental way, researchers believe. And according to the late Susan Nolen-Hoeksema, a Yale University expert on the topic, "Even when a person prone to rumination comes up with a potential solution to a significant problem, the rumination itself may induce a level of uncertainty and immobilization that makes it hard for them to move forward." Deliberately replacing unpleasant, ruminative thoughts with ones that don't cause stress is the foundation of cognitive behaviour therapy, which is one of the most effective treatments for helping people change persistently negative thought patterns. More proof that mind over matter works.

Ignoring naysayers is the kind of common sense our parents urged on us, though it's difficult to do. Criticism stings; being disliked doesn't feel good. But Linda Hasenfratz found that by treating the criticism as constructive and ignoring anything that was personal, she was able to turn it into a strength. And ultimately she proved to those naysayers that she was right for the job. By leaning away from the discomfort, she decreased her own experience of it so she could focus on what was really important to her: changing to become the kind of leader the company needed—the kind who could successfully steer it to growth and expansion.

CHAPTER FIVE

The Comfort of Control

Sometimes change isn't much of a choice. Sometimes you are blindsided by fate and your life changes irrevocably. You find out you're not just having a baby—there are triplets on the way. Or your wife announces, seemingly out of the blue, that she wants a divorce. Sometimes the discomfort of change simply can't be reframed, or repackaged in terms of gratitude, or redirected with denial. There's no sugar-coating it: you can either crawl under the covers and feel sorry for yourself, or reinvent yourself.

That's a fundamentally different situation than the one that faced Andy Puddicombe, Ray Zahab, Linda Hasenfratz, or even Jim Moss. Those people *wanted* to change to achieve a personal goal, whether that was running across a desert or running a company. But sometimes change is thrust upon you. And the discomfort that accompanies change you didn't seek and don't want is profoundly different from the discomfort associated with achieving a goal.

Jessica Watkin started figuring this out when she was just fourteen years old. "I woke up and everything was half." That's how she describes the way the world looked on the May morning in 2006 when she opened her eyes and saw nothing but blurriness out of her left eye. She assumed she'd forgotten to take out her contact lenses the night before—something a girl who's only recently started wearing them might do. So she started fishing around in her eye for the lens. It wasn't there. She began to panic. What if her parents said she wasn't responsible enough yet and made her go back to wearing glasses? It would be social death in high school—she just knew it. Jessica decided not to say anything. But as she went about her day, the blurriness on her left side didn't go away. She had to tell her mom and dad.

Alarmed, they took her to the family's optometrist, who said the blurriness was nothing to worry about; Jessica's vision was just changing, that's all. But over the next two weeks, the blurriness got even worse. Watkin, who lived in Waterloo, was referred to a specialist in London, Ontario, and he determined that her retina was detaching. The news was terrifying. She had no idea what was going on, and neither did her parents. No one they knew had ever had a detached retina—it's usually an affliction of the elderly.

Watkin was sent to Toronto, about ninety minutes away, to see a specialist at St. Michael's Hospital. She was by far the youngest person in the waiting room, and she guessed that everyone else assumed her dad was the patient. After examining her eyes, the Toronto doctor announced that her retina could only be reattached surgically. And there was even more brutal news: the surgery might not work, because Watkin had a rare genetic degenerative condition called familial exudative vitrioretinopathy (FEVR). She might lose even more sight in her left eye, and there was every possibility that her right eye could be similarly afflicted.

At the end of May, doctors removed some of the scar tissue that had developed as a result of her condition—an intricate, time-consuming surgical feat—then injected a gas into her left eye to try to hold the retina in place. After the four-hour procedure, Watkin, her left eye bandaged shut, didn't wake up for quite a while. The first thing she remembers is a nurse telling her to open her eyes. As if she even could.

Recovery was difficult. There was constant pain in the nerves around her left eye, so it felt like her eye socket was badly bruised. And there were headaches—bad ones. For weeks, she was essentially immobilized: to reduce pressure on her eye, Watkin had to spend most of her waking hours in a massage chair, the kind you see at the mall, leaning forward into a face-rest. From that position she would try to read and eat and fight off intense boredom. Her neck hurt from being in the chair all day; at night, she had to sleep face down in her bed, using the face-rest of the massage chair rather than a pillow. And every now and then, her mother had to put antibiotic drops in her eye to keep the stitches clean. That hurt almost more than anything else. Except for those brief moments, the bandage—with tape that got snarled in her hair—stayed on for days. In two weeks, as the gas that had been injected slowly dissipated, she could see some light in her left eye. But unlike the majority of retinal-detachment surgeries, hers was not successful: her retina did not reattach, and her vision did not improve. She'd been warned this was a possibility, and now there was no hiding from the fact that something was really wrong. While her doctor and her parents clung to the optimistic belief that another surgery—maybe with a slightly different technique—would help, Watkin wasn't hopeful. "After that first appointment, when they told me 'This is your condition,' I had resigned myself."

As she rode this roller coaster of emotions, the rest of the family was tested for FEVR. Her mother and four-year-old brother, Kyle, were cleared; her father and ten-year-old sister, Marissa, carried the gene linked to the disorder but so far had no symptoms. Jessica Watkin, meanwhile, remained imprisoned in her chair, with plenty of time to think about all she'd lost. There were so many things she loved to do that would now be impossible. She'd been so excited to start high school in the fall, and had imagined herself making the basketball, volleyball, or rugby team. Riding her bike? That was probably out too. What about her dream of teaching English, and maybe writing a novel? What would happen to that if she went blind? In the blink of an eye—literally—her sense of her own agency was gone. Like any teenager, she'd been trying on different hats, deciding who she wanted to be, feeling there was nothing she couldn't do. Now that feeling of boundless possibility was gone, replaced by a profound sense of uncertainty. There was no one to blame, but there wasn't any way to make sense of what was happening either. "Why me?" thoughts crossed her mind more than once. She wondered whether she'd brought her misfortune on herself in some way—whether bad karma was at play: "Oh, I yelled at Grandma that one time, [so] it must be because of that."

She felt completely isolated. Her family members were loving and supportive, of course, but they couldn't possibly understand what she was going through. She was furious about the injustice of it all. At one point, sobbing in her hospital room—fed up with what felt like endless painful tests—she'd lashed out at the eye-pressure-testing equipment. Even that went wrong: she failed to break it. Meanwhile, her father sat beside her quietly, steadily, just letting her vent.

She found that explaining what was happening to her physically helped her feel she was in control, at least of her own emotions. She could be matter-of-fact about it. "This is what I can see," she would say (mostly just light). What she didn't talk about then was how she was feeling. Instead, she wrote about it. "The hardest part for me was all the changes, the transitions at school," she remembers. "Trying to make friends. And people are really mean in high school." Never the type of girl to march up to a table full of strangers, Watkin found herself eating lunch alone. At least that was better than fielding thoughtless questions about her left eye, which was visibly scarred after surgery. Choir and drama were her saving grace; she threw herself into them and found a social niche. She figured out how to soldier on.

What she really wanted was to be like her old friends, but she knew she wasn't. Sure, they were making the same transition to high school, finding their way through a maze of cliques and boys and clubs. But her friends didn't have to miss weeks at a time for two more operations. They didn't have to teach themselves how to do linear equations because they were in the hospital when the teacher covered that part of the curriculum. They didn't have to miss gym class because their impaired vision made it dangerous. "Everyone else is dealing with buying the new thing from Abercrombie & Fitch or [deciding] what classes to take," says Watkin. "I was dealing with how I thought of myself. My . . . sense of self-worth." One challenge was that after multiple surgeries, her left eye had begun to recede into her skull, leaving a purplish bruise around her eye socket—one it wasn't safe to cover with makeup. Her eye was just too fragile. But like any teenage girl, she wanted to look pretty.

She wanted to be normal, or at least to *feel* normal—to bridge the gap between herself and her friends—so on her sixteenth birthday, she insisted on taking her driver's test. Although she was blind in her left eye, this was one rite of passage she wasn't going to miss. "I was like, 'No, I can learn with just this one eye!'" she says today, laughing at the memory. She hated driving, and "sucked at it" too, but she figured out how to stay centred in her lane—an impressive feat given her eyesight.

In grade eleven, she was given the option of getting a prosthetic eye. She'd have two matching eyes again, but she'd have to give up on the idea of ever seeing out of her left one. She was ready. And so, in the fall of 2008, alone in her hospital room, wearing the Tinker Bell nightgown her father had bought her in the Disney Store earlier that day and hugging her teddy bear, Jessica Watkin braced herself for her fourth operation, the one that would remove her left eye altogether.

Watkin has never regretted her decision, even though her prosthetic eye, which is made of coral and hand-painted by a Toronto artisan to match her other eye, has to be replaced every few years. "Now I have very matching eyes," Watkin tells me brightly. And she does. Even after talking to her at length, I would never have guessed that her left wasn't the one she was born with. Before the surgery, she used to wear her hair over her left eye, Veronica Lake–style. Afterward, her confidence grew, and she stopped hiding behind her hair. But she still felt different. Alone. She had missed a lot of school for medical appointments and procedures. Too self-conscious and shy to go to parties, she had missed out socially too. On the plus side, she did well in her courses. Yes, reading took longer with just one eye, but she managed.

By the time she was in grade twelve, she was in advanced placement English, active in the school choir, and earning top marks in drama. Still dreaming of teaching or writing a book, Watkin had decided to apply to English programs at several universities. Her sister, Marissa, was also in high school by then, and Jessica enjoyed mentoring her. Unlike their father, who remained symptom-free, Marissa had her first eye surgery at fourteen too, though her case appeared to be less severe than Jessica's. Watkin's own condition had stabilized, though, and there was no reason to believe that it would get any worse. She'd walked through fire and come out the other side only slightly singed. She had a solid group of friends, a boyfriend. Plans.

One October evening during her senior year, Watkin noticed something wasn't quite right with her functioning eye. By the end of that month, she was having trouble reading. Her right retina was detaching, but her doctor advised holding off on doing anything about it unless or until it got so bad that she couldn't function. It seemed even her doctor could only hope the worst-case scenario wouldn't come to pass. In early December, she was frantically trying to finish a term paper on Margaret Atwood's *Alias Grace* so she could get to the dress rehearsal of *Guys and Dolls*—she had a lead role, playing the part of Aunt Arvide—when she realized she couldn't really make out the words on the page anymore. "But I wasn't thinking about that," she says. "I was thinking, *Finish, get to rehearsal.*" By the end of the rehearsal, however, she knew she had no option but to pay attention to what was going on: she couldn't read her phone. Her mother was picking her up, but Watkin had no way to find her or text her. She felt as though she'd opened her eyes while swimming in a sea of oil. She could see light, but nothing else. That was on a Wednesday night, and by Friday morning she was back at St. Mike's. Her doctor scheduled her for surgery the following Thursday.

After she'd lost the vision in her left eye, Watkin had been a trouper, engaging in only the occasional act of mutiny: sneaking forbidden eyeliner (which could infect or irritate her eye), and sometimes wearing contacts instead of her glasses. Her parents would joke that they'd had a free pass through the terrible teens because Jessica had to step up and be mature. At the time, people had marvelled at her coping skills—"It's amazing that you never feel like crawling into bed and just giving up!"—and she was quietly proud of her strength too.

But the unexpected loss of sight in her "good" eye felled her, and this time she took to her bed. She stayed there for days, getting up only because she was determined not to miss the play and let her fellow actors down. So while she stayed home from school every day after the dress rehearsal, she made it to the theatre for every night of the show's four-night run. Occasionally she blundered into a prop or one of the other actors, but with their help—and the support of her boyfriend, who was also in the show—she managed to make it through the week, all the while exuding the preternatural cheer required for a musical. "It was just—keep surviving . . . Remember your lines, remember your blocking." Each night, Watkin burst into tears the moment she left the stage and removed her microphone, and she crawled right back into bed as soon as her parents helped her home.

Why bother getting up? The only thing she had to look forward to was the same surgery that had failed, three times, to restore vision in her left eye.

Watkin suffers from a rare disorder, but vision loss is far from rare in North America. Roughly 2.5 percent of the population lives with a significant degree of vision loss—more if you include often elderly people living in institutional care—and as the population ages, this number will almost certainly increase. Without question, life is more difficult for people who can't see, in ways that go beyond any challenges they have navigating a world built by and for sighted people. For one thing, they are more likely to be poor. In the United States, 30 percent of adults with vision loss between the ages of twenty-one and sixty-four live below the poverty line, and only 40 percent are employed. The picture is even bleaker in Canada, where only 30 percent of adults with vision impairment are employed and almost half live below the poverty line.

Adults with vision loss are also significantly more likely to be depressed than those with intact vision, according to research published in *JAMA Ophthalmology* in 2013. Although older people have the highest rates of depression—more than three times the norm in the general population—those aged twenty to thirty-nine aren't much better off: the research found that 13 percent were depressed, as opposed to 4.7 percent of those with no vision loss.

One study of patients with "acquired blindness" (meaning they weren't born blind) found that fully 90 percent had experienced depression. Depression and even suicidal thoughts were worst among those who were partially, rather than completely, blind. The researchers hypothesized that a completely blind patient is forced to accept the new reality and adapt. Magical thinking simply doesn't apply. Those with partial sight, however, may cling to the hope that their vision will be restored, and therefore experience feelings of loss in waves, over and over. In follow-up studies as much as four years after the onset of vision loss, half of all respondents reported that they had experienced suicidal thoughts.

Personality, apparently, plays a role in how well an individual responds to the onset of vision loss. Perhaps not surprisingly, those with a more experimental nature seem better able to adjust to the idea of themselves as blind. Those who are more timid, or more reliant on authority, seem to have more difficulty making the transition. (Incidentally, there is also research indicating that vision-impaired individuals who *regain* their sight may be at risk for depression and suicidal thoughts. Being forced to experience the world in an entirely different way, even if the development is generally considered positive, can be extraordinarily uncomfortable. Change is hard, in other words, even when that change seems to be "all good.")

Sitting in a sunlit coffee shop seven years later, Watkin, a beautiful young woman with clear blue-grey eyes and silky hair the colour of buckwheat honey, wells up, recalling how it felt at seventeen to realize that she was likely losing her sight forever. It still feels so immediate that she summons her reactions in the present tense: "I'm really worried about having kids and never seeing their faces or not watching my brother, Kyle, who's ten years younger than me, grow up. I haven't seen the world yet. And just having this realization that there's a good chance I'm never going to see London . . . One of my biggest goals when I was younger was to go to England, and I'm never going to see that."

Just before Christmas 2009, she had surgery on her right eye. Afterward, she found herself face down, once again, in a massage chair. But this time, she knew that when she got up, there was a good chance she wouldn't be able to see. She vividly remembers how, over the holidays, everyone tiptoed around her and tried to cheer her up. She hated being fawned over. Her parents moved the massage chair next to the Christmas tree so she could reach out and touch it, but she wasn't interested. Peace and goodwill weren't on her agenda.

Pretty much everything she'd known and loved—acting and choir and reading—was gone. So was her boyfriend, one month post-surgery. "It was too much, I guess," she says with a shrug. Friends did rally round, bringing her treats, putting on a *Jurassic Park* DVD, and doing their best to narrate the onscreen action. "They'd say things like, 'Now there's a dinosaur!'" she remembers, laughing. "It was an awful audio description." The sad truth, though, was that they didn't really get what her life was like anymore.

She still couldn't see anything but light, yet she didn't want to learn "those bullshit orientation tricks." She thought back longingly to the days when she could see decently out of one eye and felt a growing sense of anger about her predicament. Having no focus for her anger made it all the more difficult to bear. She knew it was absurd to get mad at her eye, but her eye was all she had to get mad at, even though "the stupid retina [didn't] listen to me."

The biggest loss of all, however, was her sense of her own identity. Suddenly, the most important thing about her wasn't that she was smart, or articulate, or could sing and act. It was now all about what she couldn't do: see.

She had been fiercely self-sufficient and independent almost to a fault—refusing to stop and ask directions, for instance, after losing the vision in her left eye. Now she was forced to rely on others all the time, even for simple tasks, like finding her way to the bathroom just after her surgery. It "sucked" having to wake up her mom if she had to pee in the night. Sometimes she stubbornly tried to find her own way to the bathroom, but often she'd blunder into something, waking her mom anyway.

In February 2010, Watkin returned to school. "I didn't want to go back," she recalls. "I was just trying to accept that this was how it was going to be. It was really hard not being able to read a book or do anything. Really depressing." Within a few months, her vision did improve slightly: she could make out vague shapes, though always with strobe-like lights flashing across her field of vision. She couldn't see objects or people straight-on, but she could make out the edges of things. And if the print was large enough and she turned her head slightly to one side or the other, she could see words on a page.

This was as good as it was going to get, apparently. She was never going to be able to see again.

Believing that she was in charge of her destiny, even as life threw her one curveball after another, made all the difference for Jessica Watkin and enabled her to persevere through, and overcome, significant physical, social, and emotional pain. The disappointment of one surgery after another didn't weigh her down with sadness; she stubbornly asserted her normality, insisting on learning to drive. Only in her senior year of high school, when it was clear that she was losing her sight altogether, did she despair. But why, given what she'd already proved to herself and the world about her determination, talent, and adaptability?

What seems to have been most difficult for her was not the loss of her eyesight but the loss of a belief that she was in control. At seventeen, Jessica Watkin was an expert at dealing with discomfort—so long as she was playing a role.

In that sense, she was very normal: a large body of research shows that feeling we have some control—before, during, and after an adverse event—significantly reduces our discomfort. Believing we have some say over how things unfold is crucial to how bearable or unbearable a blow feels.

According to Suzanne Thompson, emeritus professor of psychology at Pomona College in Claremont, California, feeling you have control over an unpleasant event takes four forms: you can believe your behaviour will influence just how unpleasant the event is; you can believe you know why the unpleasant event is happening; you can look back on the event after the fact and believe you can figure out what you did (or didn't do) that caused it; or you can simply think you have a mental strategy for dealing with the event that will make it less unpleasant. Of those four types of control, the last is the hands-down winner in terms of mitigating anxiety and physiological stress before, during, and after an uncomfortable event. Simply thinking that you have a helpful mental strategy for dealing with discomfort "appears to have uniformly positive effects on the experience of an aversive event," Thompson concluded.

Mental strategies for managing discomfort are either avoidant or non-avoidant. The trick is to figure out which one will be most helpful in a particular situation. Avoidant strategies include denying, ignoring, or distracting yourself from unpleasantness, and they make a lot of sense when you really don't have much control over a situation. Linda Hasenfratz found that simply ignoring co-workers' barbs worked beautifully. She couldn't make people stop talking, and noting and remembering each snide remark would have been a recipe for bitterness. Non-avoidant strategies include focusing on discomfort in order to better control your reactions to it, both physically and mentally. For Ray Zahab, focusing on the pain in his foot and trying to figure out what it meant allowed him to keep running. Sometimes, paying attention is useful; you can figure out the difference between training pain and serious injury. Other times, attending closely to an unpleasant event you can't alter is the first step on a slippery slope that leads to rumination, and as we've seen, catastrophizing, maximizing, and obsessing all make the subjective experience of discomfort much, much worse.

Why does simply believing you have a helpful way of thinking (or not thinking) about an event make that event more bearable? Because then you're not a helpless victim. Your thoughts and actions matter. Or at least you think they do. Studies have shown that we don't need to have any *actual* control to feel better about discomfort—we just need to *believe* we have control to reap the benefits. The illusion of control, as social scientists call it, is often adaptive and self-protective. We're better off, from a mental health perspective, believing we can withstand discomfort than we are believing we'll be crushed by it.

And as it does with pain, the meaning we assign to discomfort matters. Research shows that believing you're going through something tough in order to achieve a desirable outcome significantly reduces discomfort. For instance, Jessica Watkin believed that the surgery to remove her left eye and replace it with a prosthetic would make her look better, and that in turn helped her bounce back; when she believed that no good would come of the surgery on her right eye, she was left feeling angry and distressed.

Time and again, mind can trump matter—even when the matter is pretty scary or depressing. But for that to happen, you really do need to believe you've got some control. Sometimes, simply having information gives you a feeling of control. For instance, knowing when an uncomfortable or painful situation will start and how long it will last seems to help neutralize its sting. Researchers hypothesize that knowing when something will happen means our brains are less taxed; they don't have to be busily scanning the environment for the next threat.

But having too much information can backfire if we feel helpless. In one study of patients in a cardiac intensive care unit, researchers learned that giving them too much detailed information about their recovery didn't help. In fact, it tended to make the patients even more anxious. However, when they were not just told what to expect but also given a means of controlling the outcome—taught how to do certain exercises that would help with recovery, for instance—there was an outsized positive impact on their recovery.

Information, it seems, is a double-edged sword. If it tips us off to a problem we'd been blissfully unaware of and also can't control, it can induce anxiety. Thompson points to an episode in Los Angeles when municipal officials posted "reassuring" notices in city-run elevators urging riders to stay calm in the event they got stuck, because "there is little danger of the car dropping uncontrollably or the car running out of air." Eventually the signs had to be removed because they didn't calm riders to whom these dangers had never occurred—in fact, this "helpful" information freaked people out, transforming an uneventful elevator ride into what felt like a risky endeavour fraught with danger.

The greater the constraints and challenges, the more important it seems to be to feel that you have a modicum of control over them. In one famous study, several nursing home residents were given more control over aspects of their everyday lives. "Control" in this instance was pretty minor: they could choose which day of the week they would watch a movie, for instance, or opt to take care of a plant in their rooms. A second group of residents didn't have these options. Both groups received the same kind of help from staff, but residents in the first group were told they had a great deal of choice in their daily lives. The results were dramatic: they participated more in activities and were also happier, by their own account and also according to their nurses. During the three-week period of the study, 71 percent of the seniors who felt they had less control actually deteriorated physically, while 93 percent of those in the first group showed physical improvement. Eighteen months on, the seniors from the "in control" group had maintained those positive gains—and fewer of them had died than those in the "no control" group. Feeling we are in control of our destiny, even in small ways, really does matter, and not just as a means of helping us cope with difficulty. It can be a strong positive force in every aspect of our lives.

For Jessica Watkin, so long as discomfort was associated with positive change—"After this operation, I'll look better," or "This could help me see a little better"—she was able to power through to the other side. But when she lost sight in both her eyes, she couldn't picture the other side, nor did she believe that anything she did would change her situation. It wasn't until her discomfort was acute that the answer became obvious: she had the ultimate control—she could reinvent herself.

Even though she couldn't see, Jessica Watkin didn't yet think of herself as blind. She flat out rejected the idea of attending a school for the blind and refused to use a white cane. She felt comfortable enough getting around without one at her high school and insisted on being "normal." But there was clearly going to have to be a new normal. Until she began to change the way she perceived herself, she could not move forward.

She was assigned a vision teacher, courtesy of the school board, and learned to use a computer to type her notes at school, at a time when computers were a rare sight in the classroom. When acceptances started to come in from universities, she assessed them differently than she would have the year before. The nearest, the University of Waterloo, was her parents' first choice, not least because they worried about how she would manage and wanted her close to home. But Watkin visited the campus and rejected it. The old Jessica Watkin might have wanted to go there; the new Jessica Watkin was disappointed about how inaccessible it was. So how about her dad's alma mater, the University of Guelph, thirty minutes away from home? There she wouldn't have any family help and would have to learn how to get around all on her own. That place she loved. But to be able to go there, she would have to change her skill set—and quickly.

And so she spent the summer after grade twelve on the Guelph campus with Jeanette Dudley, an orientation and mobility expert from the Canadian National Institute for the Blind. They spent hours tracing the steps from Watkin's residence to her first class, then made their way to the cafeteria and back to her residence. Watkin still works with Dudley any time she has to make a major move to a new location. She finds the exercise both fun and very frustrating. For some reason, she retains a route best if she draws her own map—not easy, given her very limited vision.

Once classes started, her sense of control over her own life began to return. There were things she could do to make every day a little better, like learning how to organize her closets so that she'd show up at school wearing things that at least vaguely matched. She taught herself to get around off-campus using mental maps. (To make it to the café on the day we met, she guided herself by counting the three spindly trees between the subway stop and the front door.) She began to change inside too. Her memory, which was once not so good, became excellent. It had to be, so she could remember where things were and how to get around on her own.

Remarkably for a woman who despises audiobooks, she completed a four-year honours degree with a double major in English and theatre without being able to see to read. "Do you know how infuriating that is?" she asked me. It's harder to concentrate on audiobooks, and she hated having to rewind to listen to something she'd missed the first time. But instead of crawling under her covers, she gritted her teeth and figured out how to make the experience more bearable: if she kept her hands busy knitting, it was easier to stay awake while listening. So she knitted. And knitted. And knitted.

The more she felt a sense of agency returning, the more she began to view her loss of sight as something worth exploring rather than denying. Because she had always been a performer, it felt natural to do that exploration through drama. The girl who'd once yearned for "normal" had become a woman who wrote a monologue that started, "I woke up and everything was half." Not seeing could inform her vision as an artist—something she'd never imagined in those first dark days after losing the sight in her right eye.

In 2011, she owned her lack of sight in a new way, channelling her energies into leadership roles at the CNIB in Toronto. For the past six years, Watkin has worked there for five weeks each summer, leading day camps for visually impaired kids between eight and nineteen. The first summer was difficult. Everything was new: it was her first summer job, and she was still adjusting to not being able to see. But working with those kids, who'd been dealing with blindness so much longer than she had, was transformative. They'd coped with surgeries and all the other hardships that go along with visual impairment, but they did so with grace, tenacity, and optimism. "That first summer working with them settled any unrest I had about dealing with my vision loss," Watkin says today. "The campers were bright and capable and smart and wonderful and fun, and I remember each one as someone I looked up to, despite all of them being younger than [me]." Like Jim Moss, she'd decided to focus on all she had to be thankful for, which now included a sense of belonging to a community.

Through the CNIB, she met and bonded with other young visually impaired adults. In 2014, she led a CNIB program that brought ten low- or no-vision teenagers and young adults first to the lake country north of Toronto and then to the city itself. For Watkin, who was responsible for all ten, the experience was both stressful and exhilarating. Some of the participants had never been on a plane or away from their parents; some were overwhelmed by the big city and all its complexity. But what struck her most was how open they all were to the experience, despite their discomfort. Their optimism and spirit were "so rehabilitative for me," she says. She also had reasons to count her blessings: she was not completely blind, and she'd been able to go to university on her own.

Observing "the fire" in those kids, many of whom did not have the advantages she'd had, made her realize that they were all successful human beings—extraordinarily successful, given the challenges they'd faced. That insight was life-changing for her. Blindness didn't have to define her or circumscribe her life. She could be whoever she wanted to be, including the Jessica Watkin she'd once imagined. She was starting to be able to see that person very clearly in her mind's eye, which was perhaps more important in the long run than being able to see her in a mirror.

The more comfortable she felt exploring her place in the world, the more interested she became in exploring how others perceived her world. Now completing a master's degree in drama and musical theatre at the University of Toronto, Watkin is keenly interested in showing sighted people the hidden world of the blind; for one exercise, she had classmates sit in a semicircle on a darkened stage in order to show them what it's like when you can't take a performance in with your eyes and have to rely on your other senses.

For her PhD, she plans to study how performers with disabilities express themselves artistically. She also plans to learn Braille—something she's been able to put off because, in the right light and with a large enough font, she can still see letters on a page. But to earn a doctorate she must learn a second language, and Braille is a practical choice: any day, the limited vision in her right eye (currently she can see blurry shapes and the outlines of objects and faces, but has "bad peripherals" and limited depth perception) could disappear altogether. Ironically, Braille, which generally takes about two years to learn, is harder for someone like Watkin precisely because she still has some vision. She will have to focus on feeling the dots and resist the urge to look at them. "They are going to have to blindfold me," she chuckles.

Now twenty-four, she's moved past the anger, fear, and overwhelming discomfort she felt at seventeen, in part because she is not the same person. She had to reinvent herself if she didn't want to be defined by her disability or derailed by her own anger and sense of injustice. Not that she doesn't get angry about her situation from time to time. The difference now is that her anger is no longer about not being "normal," but about the ignorance "normal" people sometimes display: the barista who refuses to read a menu for her; the friend who casually asks, "Can you drive over and pick me up?" She also gets frustrated with herself if she fails to get off at her streetcar stop and arrives late at her destination as a result. But she has learned to deal with such annoyances, largely by accepting that they're natural and will never go away. She has also learned how to ask for help. Relying on others used to be anathema to her because she viewed it as weakness, but now she reminds herself every day that asking for help is okay—and sometimes it's even okay to be "a little bit bossy" to get it. The new normal includes not being embarrassed about being different. These days, Jessica Watkin doesn't mind reminding people that she can't see, and even puts up a Facebook post each year telling old high school friends not to take it personally if she doesn't bubble over with enthusiasm when they bump into each other—she's not snobby, just visually impaired, so she may not recognize them.

She still has worries, of course. Sometimes she worries that if she has kids, she may pass on her genetic condition—but when her sister (who fared slightly better in her surgeries and remains sighted) reminds her that if that happens, her kids will have a good role model, it lifts her spirits. Sometimes she worries about waking up and discovering that she has lost the trace of vision she still has in her right eye. She has nightmares about that, in fact. But as hard as that would be, she's confident she could figure out how to have a great life regardless.

After all, she's become the kind of person who does go to London, England. On her own. She used her white cane in the airport—it's one of the few places she has used it, and she did so partly to signal to others that she can't see—but ditched it to explore the city. One rainy day, she somehow wound up in an underground tunnel that clearly wasn't meant for pedestrians. She had no idea where she was. She knew she was parallel to the Thames River—she could make it out in her peripheral vision—but the stairs back up to street level had disappeared and the ones she found were unfamiliar. Utterly terrified, she forced herself to think logically and just kept walking up sets of stairs until, finally, she made out the shape of St. Paul's Cathedral, a landmark she had painstakingly mapped to get her bearings. She didn't hail a cab but made herself keep walking all the way back to her hotel, where she collapsed, exhausted. But she was also proud. It wasn't easy, but she'd stayed calm. In control.

The symbolism—believing you're lost in a dark tunnel, only to realize that somehow you've found your way out and into the light—wasn't lost on the English major. One day, Watkin hopes to write about it. But that will have to wait. These days, between her graduate degree and her activism on behalf of blind people, she's a little busy.

CHAPTER SIX

The Discomfort of Reinvention

Losing your identity when you're just beginning to find your place in the world is disorienting, as Jessica Watkin discovered. But let's say your identity was virtually preordained at birth and you've lived up to it and then some, earning stripes for your mastery and rising to the top of your field. Then you wake up one morning and it becomes clear that the rules of your world have shifted, almost overnight. Moving on from something that has worked for you for a long time can be one of the most challenging—and uncomfortable—forms of change. Everything you've mastered is now obsolete. If you stubbornly cling to the old ways, you will almost certainly fail. That's true for a weight-loss regimen that no longer works for you, and it's true for changing social norms. But what if that failure to evolve could result in human carnage on an unimaginable scale? In that case, reinvention would not be optional. It would be a moral imperative. This was the situation faced by Stanley McChrystal in Iraq in 2003.

Now sixty-two, McChrystal carries himself with an almost regal bearing. Though retired, he still looks every inch the four-star general. His back is arrow-straight, and every move he makes is quick and purposeful. He has the ascetic look and lean physique of a man who runs seven miles a day but eats only one meal, and a whiff of old-school courtesy about him. But McChrystal is not a typical military man. Best known for his candour and rebellious streak, he built his name on disrupting the status quo and reinventing the military. The fact that those same traits occasioned his fall from grace, forcing him to reinvent himself, makes him something of an expert on the discomfort that accompanies change.

McChrystal cemented his reputation during the mid-2000s, when he was in charge of the US military's Joint Special Operations Command (JSOC). JSOC oversees covert military operations, counterterrorism reconnaissance in hostile areas, and special intelligence missions. Special forces such as the Navy's SEAL Team Six and the Army's Delta Force are under its command, and JSOC has established a reputation for training the best of the best and dispatching them to deal with the worst of the worst. It's also been called "the most secretive force" in the US military, home to the "snake-eating, slit-their-throats 'black ops' guys."

Between 2003 and 2008, McChrystal directed some of the Iraq War's most pivotal operations, including the December 2003 capture of Saddam Hussein. But when he arrived in Iraq that fall to head up Task Force 714, the counterterrorism unit of the JSOC, success was far from assured. As he recounts in his book *Team of Teams*, what began as a "heavily conventional military campaign to unseat the regime of Saddam Hussein had, by the fall of 2003, become a bitter, unconventional struggle against frustrated Sunnis who increasingly coalesced around a charismatic Jordanian extremist [named] Abu Musab al-Zarqawi"—a struggle that "was as confounding as it was bloody." Why confounding? Because the US military in general, and the task force in particular, seemed to have every possible advantage over its enemy: lavish resources, sophisticated equipment, and huge numbers of "exquisitely trained" soldiers. But al-Qaeda was kicking the task force's collective butt. How could it be that a ragtag group of extremists was able to outgun the US military's best?

Nothing in McChrystal's own training or experience had prepared him for an enemy like al-Qaeda. For one thing, the extremists were so certain God was on their side that they were willing to blow themselves up for their cause. For another, new information technologies allowed them to operate in a decentralized fashion, so they could move rapidly, strike with devastating force, and then disappear, fading back into the local population. They could communicate in ways that allowed them to "flow a little bit like water," as McChrystal put it to me—ways that traditional organizations both "couldn't and needed to face."

Al-Qaeda had, in other words, changed the rules of engagement, but the Americans were still using the same old playbook. The US military's first big problem was its organizational structure, which, like that of most traditional businesses, was strictly hierarchical. Information was sent up through the ranks so decisions could be made at the top and then relayed back down to operatives in the field—a pokey, laborious process that was further complicated by time-consuming accountability protocols.

Al-Qaeda's operations, meanwhile, much more closely resembled those of a disruptive business start-up. Instead of being a rigid hierarchy, the organization was made up of loose cells of fighters, and they used technology to communicate information and changes of plan far more rapidly. Unhampered by a traditional chain of communications, al-Qaeda fighters could exchange information and act on it in near real time. And anyone who could get the job done was empowered to act.

The JSOC—which had been created to respond to legacy terrorist threats after the Americans' disastrous attempt to rescue hostages in Iran in 1980—suddenly found itself in a situation that, as McChrystal writes in *Team of Teams*, "had more in common with a Fortune 500 company trying to fight off a swarm of start-ups than [it] did with battling Nazi Germany in World War II." The Americans' traditional reliance on studying an enemy's past behavioural patterns to predict its future moves was simply not working anymore. The patterns seemed to be forever changing, and in any event were utterly confusing. Even the knowledge gained from guerrilla warfare in Vietnam proved to be of little value against this new enemy, according to McChrystal. Although the Viet Cong's organizational structure was less rigid than the American military's, there was still *some* structure. For al-Qaeda, and subsequent iterations of terrorist groups, the association between the leadership and its members could be extremely tenuous, so patterns of behaviour were that much harder to predict.

In 2004, McChrystal and the rest of the JSOC leadership were only just beginning to understand what they were up against. Accustomed to conducting occasional "exquisitely planned surgical strikes," the counterterrorism experts initially tried simply to pick up the pace, ordering daily strikes and, often, multiple simultaneous raids. Within a few months, they figured they'd built an "awesome machine." However, it quickly became depressingly clear that merely increasing the speed and magnitude of their operation wasn't going to turn the tide. If they hoped to contain and ultimately reduce the threat posed by al-Qaeda in Iraq, the members of the task force—and ultimately the JSOC—would have to reinvent the way they operated. And they'd have to do it on the fly, embarking on an across-the-board restructuring of operations while they were waging a bloody war.

The urgency of the situation in Iraq—in 2003 alone, 586 American soldiers were killed, and there were thousands of civilian casualties—helped spur the US military to take action. Task Force 714 didn't have the luxury of time to study the problem. Americans were dying, and support for the war was waning at home (while a majority had backed the war when it began in 2003, by 2004 Americans were increasingly divided, with the majority telling pollsters it had been a mistake to send troops to Iraq). As difficult as it was to concede that time-tested methods weren't working, there was no other choice.

Something had to change. Fast.

• • • •

McChrystal's background and personality made him a natural to shake up a moribund organization. The son and grandson of highly decorated soldiers, he seemed destined to wind up in the family business. His father—a major general in the army who'd served in Korea and Vietnam, earning four Silver Stars, the Bronze Star, the Distinguished Service Medal, and a Combat Infantryman Badge—was his idol. All four of his brothers were military men, and his sister married one. His mother, who died suddenly when he was a junior in high school, also shaped the soldier he would become. She was, he writes in his 2012 memoir, *My Share of the Task*, a woman of "extraordinary energy" and a passionate reader who introduced him to Tennyson, Greek and Roman mythology, and tales of the Scottish chiefs. Thanks to her, he grew up steeped in stories of Roland at the pass and Horatius at the bridge; thanks to his father, there was a real live hero on the premises (when he wasn't away in the trenches).

But although McChrystal viewed the military as his calling, he was not the kind of kid who got up every morning and made sure the hospital corners on his bed were perfect. He had an attitude. In July 1972, when he followed in his father's footsteps and entered West Point, he couldn't see the purpose of standing in the hot sun with the other cadets while a colonel lectured them about securing their collars with the wire stays they'd been issued. McChrystal, all of eighteen, was incredulous. He'd come to West Point to learn about leadership and courage, not collar stays. He must have communicated his displeasure, because he was called on the carpet countless times for infractions. (Among them, he was caught drinking in his room and also laughing at an upperclassman who reprimanded him for another offence.) By the end of his first year, he'd logged 127 hours in "the Area," a paved courtyard cadets had to walk around as punishment, and as he wrote in his memoirs, he'd come to view West Point as a "penal colony."

Still, he was never censured for dishonourable conduct. And despite his penchant for challenging authority and his less-than-stellar academic record—math and science were not his forte—he had boundless curiosity and was considered brilliant and something of a Renaissance man. As the managing editor of West Point's literary magazine, he also wrote short stories.

In the spring of his second year, McChrystal took his first step toward becoming what he considered to be a real soldier, competing to be one of the handful of cadets allowed to attend Ranger School during the summer break. The gruelling sixty-one-day course, created at the start of the Korean War to simulate the stress of combat, was then and is still one of the US military's most brutal training programs. Designed to test physical stamina and mental toughness, it involves three phases (in McChrystal's day, they were called the "crawl," "walk," and "run"), during which soldiers traverse woods, mountains, and swamps and are pushed to the limits not just physically but also mentally and emotionally. Typically, only half of those who attempt the program complete even the first of its three phases.

The legendary hardships of Ranger School—sleep deprivation, hunger, physical exhaustion, instructors who delight in making cadets' lives a living hell—appealed to

McChrystal. He wanted to earn the tab that would adorn the left shoulder of his uniform and establish his bona fides as a soldier.

But he didn't make the cut. He didn't have the grades or the physical training scores, and his disciplinary history made his record look even worse. The rejection was a blow. It was also a wake-up call: if he wanted to be a real soldier, he had to start acting like one. He had to become the kind of guy who followed rules, instead of questioning or mocking them. Chastened, McChrystal returned for his third year determined to change.

With his record, reinventing himself would not be easy. But as fate would have it, that year a new officer, Major David Baratto, was in charge of the cadets, and he'd scheduled counselling sessions with those under his command. McChrystal braced himself for a stern lecture. Instead, Major Baratto declared that he thought the mediocre student with a chip on his shoulder was destined to become a great leader.

McChrystal was stunned. What the . . . ? Apparently Baratto was a glass-half-full type, the sort who could give the ever-optimistic Jim Moss a run for his money, so he'd chosen to focus not on grades or disciplinary infractions but on McChrystal's peer leadership ratings, which were excellent. Classmates looked up to him, jokingly comparing him to Captain Virgil Hilts, the irreverent, rebellious American soldier played by Steve McQueen in *The Great Escape*. In the movie, Hilts spends his time in a Second World War German POW camp repeatedly trying to escape and repeatedly getting caught and sent to the "cooler," or solitary, from whose confines he promptly plots his next breakout. McChrystal's peers thought he had that same cool confidence and dogged brand of courage. And Baratto thought those qualities, if harnessed and channelled properly, could make him a first-rate leader.

Baratto's faith in McChrystal proved to be a turning point in his life. By treating the younger man as a future officer instead of a wayward cadet who had to be lectured about collar stays, the major wisely appealed to McChrystal's better instincts. Bolstered by his officer's confidence in him, tired of being disciplined, and determined to show what he could do, McChrystal buckled down and resolved to match his "professional drive with personal focus." His grades improved and so did his military performance. His class rank leapt meteorically, to 298 out of 834. Like his father and grandfather before him, he chose to enter the infantry, and on graduation day, June 2, 1976, his father commissioned him as a second lieutenant. Stan McChrystal had fallen into line, reinventing himself as a true military man. Today, looking back at the young cadet who wrestled with authority every step of the way, all he can do is laugh. "Authority won," he tells me, grinning.

But that's not the whole story. It would be more accurate to say that while he developed real respect for authority, he never really stopped questioning it. In November 1976, he finally started Ranger School training, and during that nine-week course, he learned the difference between being a rebel without a cause and rebelling for a worthy cause.

One night, after his group had just completed a six-mile speed march, their instructor led them to a training field and ordered them to crawl through an icy, slush-filled pit and navigate an obstacle course. It was dark, cold, and the men were already exhausted and chilled with sweat from their march. As they tried to traverse the course, their breathing came in gasps and they could no longer grasp the ropes to pull themselves along. In his memoir, McChrystal recounts feeling that a line had been crossed: this wasn't just tough—it was "dangerously stupid." Just then, the field lights flashed and a lower-ranked officer shouted for them to stop immediately and return to their barracks. Their own officer, his feathers ruffled, protested. How dare a subordinate issue a countermanding order? But it was too late. The men were already running from the field. That junior officer's courage in overruling a superior to "stop the foolishness" and protect the men made a lasting impression on McChrystal. The lesson was driven home to him even more forcefully several weeks later, when two Ranger students died from exposure on a similar exercise.

A decade later, by which time he was a regimental commander of the 3rd Ranger Battalion, McChrystal had the opportunity to "stop the foolishness" himself. In the eighties, almost two dozen Rangers were killed during training accidents; he set out to revolutionize their training regime. It wasn't the last time he'd seek to transform a system he considered outdated.

Once McChrystal and his team realized that success in Iraq would require the wholesale reinvention of operations, the question was how to achieve that. Task Force 714 was part of a large, institutionalized, disciplined military machine and a "veritable leviathan" compared to al-Qaeda.

The Americans' only shot at responding to the complex threats coming at them every day was to toss a century of conventional wisdom out the window and retool from the ground up. So under McChrystal's leadership, that's exactly what they did. They got rid of familiar organizational structures originally established to foster efficiency and installed others designed to create a culture of "organic fluidity."

Two sacrosanct principles guided the transformation: the new and improved task force had to have an utterly transparent information-sharing process (what McChrystal called "shared consciousness"), and all decision-making authority had to be decentralized (which he dubbed "empowered execution"). In order to democratize the flow of information, they tore down walls that had been erected to foster efficiency or protect classified information. Literally.

To start, the task force radically reconfigured its workspace at Balad Air Base, north of Baghdad, doing away with all offices and cubicles in favour of an open-concept plan with a mission-control area in the middle. All commanders and key intelligence personnel sat in front of a circular bank of computer screens to monitor ongoing missions. Like spokes on a wheel, the other members of the team radiated out from that central hub. The new workspace layout meant that groups within the task force no longer received information solely about their own missions; they heard about everyone else's too. Suddenly, information that had once been deemed relevant only to the senior officer in charge of a particular unit became available to everyone.

McChrystal designated the entire room a top-security space, meaning that just about anything related to a mission could be discussed openly. He took almost all calls on speakerphone, so anyone who cared to could listen in—even to calls about the most sensitive operations. "This could make people uncomfortable, sometimes intensely so," McChrystal writes. But never once, he says, did he see a case where openness hurt more than it helped. Similarly, previously top-secret video conversations suddenly included a wide range of people in Iraq and Washington—something McChrystal knew could increase the potential for leaks to the media. But he didn't care, nor was he one to temper his language: "I had no interest in, and we had no time for, painting a rosy picture of what was in reality a hellish scene."

In describing this new dynamic, McChrystal told me it created an "incredible kind of lateral transparency. Instead of information going up and down the verticals as [it] traditionally had, we democratized it. Now when you think about that, you're certainly opening Pandora's box to some complications. But suddenly if you hear what people [are saying] in other parts of the organization, you start to get a sense of what they're doing. And then you start developing lateral connections to them, [and] that's where trust comes."

Trust was the key to building the relationships required by this "team of teams" approach. At least one person on each team was expected to know—and trust—someone on another team. The idea was to create a network of empathy so that people would trust each another to share information. McChrystal worked hard to foster openness across the whole organization and create an atmosphere where subordinates were required to be candid—and so were leaders. This meant approaching other team members as human beings first, and military men and women second. "There's an art to asking questions as a senior leader that can get candour back," he explained to me. "If you come to me and say, 'How are you this morning?' my response is 'Fine.' Because I think you really are just making conversation. But if you really stop and you say, 'No, tell me about how you are or what's going on in your life,' then it's a different conversation."

Meanwhile, McChrystal and his cohorts put their smallest units under the microscope to determine which of their behaviours worked best, then found ways to extend those behaviours across three continents. The strategy effectively turned the JSOC's four-thousand-person organization into a team of teams that operated with the kind of nimbleness and flexibility normally achievable only on a smaller scale. Groups that had been low down on the totem pole were given more autonomy and encouraged to experiment and share what they'd learned across the entire organization; technology was used to foster a sense of common purpose in a way that would have been unimaginable a decade earlier. "Almost everything we did ran against the grain of military tradition and organizational practice," McChrystal explained. "We abandoned many of the precepts that had helped establish our efficacy in the twentieth century, because the twenty-first century is a different game with different rules."

Remarkably, very little of this transformation was actually planned—and very few of the plans that were developed unfolded as expected. Operating largely on instinct—and whenever possible drawing on insights gained from hard-won experience—McChrystal and his senior leaders assessed and adapted, assessed and adapted, transforming the organization on the fly. Iteration by iteration, they "morphed, and morphed again."

But you don't turn things upside down to the extent that McChrystal did without putting a few noses out of joint. Not surprisingly, he encountered resistance. Sharing information without overthinking protocol was central to the plan, but intelligence agencies such as the CIA and FBI were wary of divulging information to people outside their own walls, and they weren't even accustomed to sharing it with each other. To them, McChrystal writes, "Our way of work was anathema." One agency, for the first full year of the experiment, offered the same thing every day: "Nothing new to report on our end."

The soldiers under McChrystal's command also had to wrestle with discomfort. If you'd been trained as an engineer at West Point, wrapping your head around the idea that you're never going to find a perfect solution—just a different one that works on different days—was "fundamentally disturbing." And what about all those newly empowered people? Some were also newly uneasy with their increased level of responsibility. It's harder to blame someone else (including your boss) for foul-ups when the buck stops with you.

One thing McChrystal had going for him was that those below him were less likely to question authority than he was. Ironically, the respect others had for tradition and hierarchy may have helped him introduce sweeping changes designed to break down the traditional hierarchical approach that was hampering the military in Iraq. Regardless, he recognized that change is as difficult for those who follow orders as it is for those prone to questioning orders.

"Change is painful," McChrystal later told the *Washington Post*. "And people are always reticent to accept a lot of pain. If you come in with a bulldozer mentality and say, 'What you are and what you do is wrong,' you're not apt to get a long line of people following you. But if you make the case for it, and you show people that the status quo is unsustainable, then I find that you get a lot further." He convinced people to climb aboard by acknowledging how difficult it would be for all of them. "You have to have empathy. In an organization that is unwilling to change, find the opportunity to talk and interact with people—figure out why they don't want to change. It could be habits, it could be [that] people's personal . . . reputations are defined by the role they're in."

Meanwhile, even though he was the one rocking the boat, he too sometimes found it hard to change. Having grown up in the military and risen to the top within its confines, he had to overcome his natural inclination to stick with what had worked for him. It wasn't easy to throw out a playbook he'd spent decades studying and essentially unlearn much of what he thought he knew about the way war worked.

He anchored himself by keeping a firm eye on his goals: decentralizing decision-making authority, ensuring transparent communications, giving small groups the freedom to try and fail, encouraging everyone to share what they were learning with others, determining which practices were succeeding in the smallest units and implementing those measures across the entire organization. And most of all, empowering those closest to the problems to act on them, decisively.

What kept him going through all the upheaval? "The thing that really drives me is that I hate to lose," he told me. "So the only thing that really forced me to change was not an intellectual argument. It was the absolute realization that if I didn't change or my organization didn't change, we would lose."

It's hard to imagine the kind of stress McChrystal was under: trying to turn the *Titanic* around, manage complex internal politics, rally the troops, keep his soldiers out of harm's way, and win what must often have seemed like an unwinnable war. Fortunately, his personal history and his training had helped him develop a skill set that enabled him to manage an extraordinary level of stress.

Stress affects each of us differently. How we react to it depends on our genes, history, spiritual beliefs, fitness level, support systems—and also how our brains are wired. How our brains react under duress is the province of Dr. Martin Paulus, a psychiatrist with the US Navy's Warfighter Performance Department in San Diego, California. Dr. Paulus wants to understand why some people's brains perform better under stress to see if it's possible to develop "workouts" targeted to the specific part of the brain that manages our stress responses, with a view to improving its performance. To figure this out, he gets to study some of the most stress-resistant brains on earth: those belonging to Navy SEALs.

To join the elite ranks of the SEALs (an abbreviation of "sea," "air," and "land"), recruits must endure weeks of punishing physical, emotional, and psychological ordeals, during which they're cold, wet, hungry, sandy, and sleep-deprived. In one exercise designed to simulate the feeling of drowning (and teach recruits how to control their fear in a life-threatening situation), SEALs jump into a pool wearing goggles and an oxygen tank. Instructors then tear off their goggles, rip their regulators from their mouths, disconnect their oxygen lines, and spin them around until their lungs are bursting and they're totally disoriented. Throughout, they have to stay focused and remain calm.

Each year approximately one thousand recruits sign up for this kind of training, and about a quarter of those complete the program. Paulus and his colleagues study the brains of those who make the cut. For one study, researchers scanned the brains of eleven off-duty SEALs and twenty-three healthy "normal" male volunteers while they performed an "emotion face-processing task." In this exercise, images of angry, fearful, or happy faces flashed at the top of the subjects' computer screens at five-second intervals, six times in a row. Two faces displaying different emotional expressions appeared at the bottom of their screens. By pressing keys on a button box, the subjects had to match the faces at the bottom with the face at the top that had the same emotional expression. Dr. Paulus and his team discovered that SEALs were quicker to react to negative signals, and when they did, a part of their brains called the insula lit up more. The insula is tucked deep in the brain's cerebral cortex and is central to how we experience the world. It's associated with our ability to sense pain, experience emotion, and anticipate and navigate red-alert situations.

The study had a relatively small sample size, and it didn't address whether the observed neural differences were pre-existing or a consequence of SEAL training. But the research did clarify that the brain circuitry of elite fighters "deployed greater resources when it perceived a potential threat-related facial expression—and fewer resources when no threat existed in the facial expression." In other words, SEALs' brains know better than yours or mine when a threat is imminent and it's time to marshal resources to fight it—and also when to conserve energy because no pressing danger is at hand. To put it another way, SEALs know when it's worthwhile to be stressed, so in fact they may be less stressed overall than the rest of us. Their neural networks aren't constantly jangling in response to minor problems, like bad traffic or an argument at home. This may go a long way toward explaining how Stan McChrystal coped in Iraq: his military training, especially with the Rangers, probably helped wire his brain to deal exceptionally well with stress.

Paulus believes that it's possible to boost the performance of brains that don't perform at Navy SEAL levels by rewiring their neural networks. The thing to remember about stress responses, he cautions, is that "they kick in no matter what the stressor is—whether you're having a family problem or your buddy just got blown up. The brain only has so many ways of organizing a reasonable response." The aim, then, is to teach brains a SEAL-like trick: don't over-react when the problem is minor.

Fortunately, there's evidence that you don't need to spend years training as a commando to optimize your brain's capacity to cope with stress. It seems that an eight-week mindfulness course, of the sort created by Jon Kabat-Zinn, will do just fine. Dr. Paulus also headed up a 2014 study—the results of which were published in *The American Journal of Psychiatry*—in which researchers found that when Marines received mindfulness training, their heart and breathing rates returned to baseline levels faster after an intense training session simulating battlefield conditions than those of Marines who didn't receive the mindfulness instruction. Their sleep quality also improved, and blood tests suggested that their immune function received a boost as well.

As the rates of physical and mental injuries, suicides, and divorces climbed ever higher in the 2000s, owing to increasing numbers of troops serving multiple deployments, the US military began pumping massive resources into exploring psychological health. Harnessing the latest neuroscientific tools (such as brain scans), as well as non-invasive alternative techniques (such as meditation), they began looking in earnest for ways to inoculate soldiers' brains against the worst effects of stress.

Mindfulness training may be the silver bullet they've been seeking. According to a study published in 2015, mindfulness training not only helps soldiers cope with stress and anxiety, but also keeps them alert and focused even when they're under tremendous pressure. Led by psychologist Amishi Jha of the University of Miami, the study looked at eighty American soldiers preparing for a counter-insurgency combat deployment to Afghanistan. Previous studies had shown that the months and weeks pre-deployment, when soldiers are busily re-upping their skills while preparing to leave their families, are especially stressful. Jha had already conducted research showing that mindfulness-based training protected against degradation in working memory—or the part of our cognitive system that processes information and guides decision making—in pre-deployment Marines.

This time, to see if mindfulness might also help keep soldiers focused and alert, Jha and her team of researchers divided the eighty subjects into two groups, both of which received an hour a week of training for eight weeks. Those in the first group attended lectures on mindfulness, while those in the second group actually practised exercises in class and discussed their experiences; all eighty soldiers were told to do mindfulness exercises at home and to log the amount of time they spent on them. In addition, there were two control groups—the first made up of Marine Corps reservists preparing to deploy to Iraq, and the second composed of civilians. Neither of the control groups received any training at all. At the beginning and the end of the eight-week period, all subjects took a test that measures sustained attention: they were asked to sit in front of a computer and press the space bar each time a particular number was displayed on the screen.

As the researchers noted, although soldiers are trained to stand at attention, their minds are not. Soldiers are human too, and so they daydream and get distracted. In fact, their thoughts (and ours) are wandering 30 to 50 percent of the time they're awake—possibly even more, because people generally underestimate the amount of time they spend "off-task." (In other words, they report much less mind-wandering time than is actually measured when they're hooked up to brain sensors—which is only natural, because when we're wandering we are least aware of our own mental processes.) When we're off-task, the researchers explained, our "attentional resources" are essentially hijacked, "commandeered by internally generated thought"—a significant problem if you're a soldier and your survival depends on situational awareness. And there's worse news: attentional lapses tend to increase when you're under the greatest stress. In other words, when the stakes are highest and soldiers really need to pay close attention to be able to adjust their responses in real time, their minds are most likely to wander.

Which is exactly what seemed to happen to the soldiers in this study who hadn't learned about mindfulness. Those in the no-training group did much worse on the attention test at the end of the experiment, when deployment was imminent, than they had at the beginning; it was more difficult for them to sustain their attention and they were less accurate. The civilians who'd received no training, on the other hand, did about the same both times—they weren't getting ready to go to war.

Mindfulness training clearly helped soldiers' brains "stand at attention." Those who'd been trained via lectures experienced less deterioration in their scores than those in the control group, while those who'd trained by practising in class experienced almost no decrease at all over the eight-week period when they were getting ready to go to Afghanistan. Their brains were better able to focus and stay on task, despite the stress they were under. Jha and her researchers believe that practising mindfulness also increases soldiers' resilience, improving their ability to bounce back from high-stress episodes. Clearly, just tuning into the moment, without judgment, helps alleviate a lot of the discomfort associated with change—even if that change involves going into combat.

Whether Stan McChrystal would call it mindfulness or not, he wound up practising something similar in Iraq, where he was forced to really observe what was happening, rather than respond reflexively, in order to change it. It was only when he stopped being mindful that he really ran into the kind of difficulty that would force the most challenging reinvention he's undertaken yet: his own.

It wouldn't be a stretch to say that three years after McChrystal arrived in Baghdad, the Americans found themselves spinning around underwater without regulators or oxygen tanks. By 2006, according to the *New York Times*, the situation in Iraq had descended into "cataclysm" and the country was in the throes of a bloody civil war. "A thousand civilians were dying every month" because al-Qaeda gunmen and suicide bombers were massacring Shiite civilians, and the Shiite militias were massacring young Sunni men.

By the end of 2007, after McChrystal's commandos conducted a series of operations, the situation had changed dramatically. Many reputable news organizations, typically stingy with praise when it comes to military invasions, lauded McChrystal's achievements in Iraq. CNN's national security analyst credited him with transforming and modernizing the JSOC into a "force of unprecedented agility and lethality." *Esquire* echoed the compliment and specifically praised the June 2006 targeted killing of Abu Musab al-Zarqawi (who'd headed al-Qaeda in Iraq since 2004) "using a virtuoso combination of signals intelligence, . . . human intelligence, . . . interrogation (the best of the best, drawn from government, the military, and the private sector), and special ops." The magazine also quoted an ex-Ranger, whom it identified as a leading expert on counter-insurgency. He argued that the turnaround was accomplished not by the 2007 deployment of more than twenty thousand additional soldiers, but rather "by the elite killers of JSOC—as led by Stan McChrystal."

The impact of his leadership can be measured in other ways too. In December 2006, there were more than 140 suicide bombings in Baghdad; in December 2007, there were just five. According to one retired lieutenant colonel of the elite British Special Air Service, "General McChrystal delivered that statistic." He raised the "hard, nasty business" of counterterrorism—of "black ops"—to an industrial scale, with ten nightly raids throughout the city, or three hundred a month. And apparently, the general regularly joined these raids himself. Then there was his openness to bringing in a new kind of operative: the computer geek. *Rolling Stone* reported that McChrystal "systematically mapped out terrorist networks, targeting specific insurgents and hunting them down—often with the help of cyberfreaks traditionally shunned by the military." A special forces commando who worked with McChrystal in Iraq told a reporter, "The Boss would find the 24-year-old kid with a nose ring, with some fucking brilliant degree from MIT, sitting in the corner with 16 computer monitors humming . . . [and] say, 'Hey—you fucking muscleheads couldn't find lunch without help. You got to work together with these guys.'"

McChrystal's record in Iraq was not unblemished. According to a report released by Human Rights Watch in 2006, detainees at Camp Nama—a unit under his command—"were regularly stripped naked, subjected to sleep deprivation and extreme cold, placed in painful stress positions, and beaten." McChrystal was also criticized for personally signing off on a Silver Star recommendation for former NFL football player Pat Tillman, who was killed in a friendly fire incident in Afghanistan in April 2004. Days after Tillman's death, McChrystal sent an urgent memo to Washington warning that President Bush should avoid making any public statements about the cause of Tillman's death, lest the circumstances surrounding it became widely known and cause Bush "public embarrassment." (Tillman's own mother didn't learn that her son had been killed by friendly fire until March 2007, when the memo was leaked.)

Nevertheless, McChrystal was widely considered to have succeeded in Iraq, and in June 2009, he was promoted to lead the US forces in Afghanistan, a post he held for a little over a year. And then his candour—so key to his success in creating the team of teams in Iraq—became his undoing. McChrystal and several aides made mocking remarks about President Obama and other Washington brass to a reporter from *Rolling Stone*, and in the subsequent article, he and his team came across as having a disregard for the chain of command. This time, he'd tossed the wrong rulebook.

The article engendered fury in Washington. Suddenly, publications like the *New Yorker* were lambasting the fearless commander for showing "contempt for political and diplomatic processes." The *Wall Street Journal* reported that even some of his strongest advocates failed to come to his defence. Long inured to discomfort, McChrystal had apparently failed to appreciate how uncomfortable his remarks would make others. Shortly after the article was published, he was summoned to Washington. He resigned his post in Afghanistan and announced his retirement days later. His military career was over.

For a man who'd spent a good portion of his career operating in secrecy—and winning kudos for his conduct—it was disconcerting to be at the centre of a media firestorm and see himself portrayed disparagingly on the news. He later said that as commander, he took responsibility for the comments in the article, even though he didn't view them as insubordinate at the time and believed the magazine had depicted his team unfairly. He regretted how events had unfolded, mainly because he could no longer carry out a mission about which he cared deeply—and one that the soldiers he'd brought to Afghanistan had trusted him to execute. He felt he'd let his people down.

And there was more to come. After the dust settled, he found himself staring down the barrel of another reinvention he'd never wanted to engineer: he'd fallen from grace in a way that threatened to erase all his previous accomplishments. If he didn't want the words "dissed the commander-in-chief" in his obituary, he was going to have to find a way to redeem himself. But every familiar road to redemption was now closed to him. He'd been a soldier practically since he started shaving; the military was all he knew.

Four years later, he confessed in a LinkedIn blog post that when he left the profession that had been his life's focus, his world "suddenly and profoundly" changed, and his first inclination was to feel sorry for himself. "I've never been more tempted to feel like a victim," he wrote. After a lifetime of success, public humiliation must have been a searing trial. The desire to rant to pundits or write a tell-all, McChrystal admitted, was overpowering. But with his wife's support, he made a deliberate decision not to define himself through his departure, or let others do so either.

When discomfort is acute, it can trigger a fight-or-flight instinct, but for McChrystal—and Jessica Watkin—it never felt right to hide under the covers or unleash his anger at the world (or himself). That didn't provide a way forward. Instead, discomfort became a prompt to look inward and ask some very mindful questions: Who am I, now that the place I once occupied in the world is closed to me forever? And who would I like to be?

When change occasions a loss of identity, whether that identity is "greatest fighting force in the world" or "respected four-star general," reinvention is the path that provides the greatest sense of control, as Jessica Watkin found. For McChrystal, it meant reinventing himself as a guy who doesn't throw out the rulebook anymore—now he writes it, teaching others to serve and lead. To that end, he's taught courses in international relations at Yale, written two books, and launched an eponymous consulting group whose mission is to bring the lessons of the battlefield to the boardroom. The McChrystal Group advises businesses in need of adaptation on how to transform their operations and improve their performance by using the "team of teams" approach its founder developed in Iraq.

One of the lessons McChrystal now teaches others is that change—the most awful and personal kind, which you don't see coming and which leaves you reeling—is inevitable. "Everybody in life is going to have something unexpected happen to him," he told a reporter in 2015. "It can be as shocking as coming home to find your spouse with another person. It can be being let go suddenly from your position. You're going to have something that shifts the ground under your feet."

Coping with stress, so you stay calm and grounded even when your life is blowing up, is essential. You need to be able to focus on what's really important and sort out the real threats from the mere distractions. McChrystal, of course, had had a lifetime of training to face just such a situation, and the skills he developed as a soldier were the ones that helped him most when the life he needed to save, and remake, was his own.

Every rebuilding project, he observed, starts the same way: "First, you're going to look inward and say, 'Who am I?' And if who you were was entirely based upon the position you were in, or the headlines you got in the newspaper, or you had essentially subcontracted out your self-worth to the judgments of others, then you're going to be like tumbleweed. You're going to be blown." If, however, your identity is based on some core skills and strengths, you will figure out a new way to capitalize on them. The fact that McChrystal has been able to do just that, embarking on a successful second act so soon after act one ended in flames, speaks volumes for his degree of comfort with change. He'd built his career around it—so when that career ended abruptly, he knew what he had to do.

McChrystal believes that hard-wiring people to deal with change is the challenge of the future, both in the military and in business. Leaders wedded to a single solution will fail, he says; the problems will be different tomorrow (and the day after tomorrow), and so will the solutions. As he writes in *Team of Teams*, there is "comfort in 'doubling down' on proven processes, regardless of their efficacy. Few of us are criticized if we faithfully do what has worked many times before. But feeling comfortable or dodging criticism should not be our measure of success. There's likely a place in paradise for people who tried hard, but what really matters is succeeding. If that requires you to change, that's your mission."

CHAPTER SEVEN

A Majority of One

Ten milliseconds faster and Jennifer Heil would have won a bronze medal in freestyle skiing at her first-ever Olympics in Salt Lake City in 2002. That's less time than it takes a hummingbird to flutter its wings. But for Heil, it might as well have been an eternity. It meant the teenager from Spruce Grove, Alberta, wouldn't be on the podium.

At just eighteen—the youngest member of the 2002 Canadian Olympic team—the slight blond skier with the toothy grin had been eager to prove herself. Her intense drive was one of the reasons her teammates called her Little Pepper (the other was her height: five foot three). Losing by one one-hundredth of a point wasn't the outcome she'd hoped for, but she had reason for pride. She'd proved herself after all.

Only she hadn't. The real test hadn't yet happened. It was still ahead of her, and it wouldn't occur in a competition or even on the slopes. Instead, it happened when she decided to change everything in her life: leave home, turn her back on the skiing establishment, trust her own instincts, and reinvent herself—then start a business to teach others how to do the same.

Heil has been on skis since she was two years old. Both her parents loved skiing, but her father, a lawyer, was especially passionate. "I think it was his goal to have me ski a double black diamond run by time I was four!" she has said, only half-jokingly. By the time she was in elementary school, she was spending just about every weekend during ski season on the slopes. Most Friday nights, Heil and her older sister, Amie, piled into the back seat of the car for the four-hour drive to the ski hill, catching up on their homework—and their sleep—during the ride home on Sunday. Heil's parents used to pull the girls out of school every year for family trips to ski resorts outside Alberta too, and she quickly came to associate skiing with family fun, excitement, and adventure.

An active kid from the get-go—moving "always felt like a place of freedom," she says—Heil took dance classes, gymnastics, and swimming, and she and her friends were constantly biking or rollerblading around their small community west of Edmonton. She denies being a tomboy but then, laughing, recalls how her mother had to "wrestle her into a dress" occasionally. "So maybe 'tomboy' is fair," she admits. One thing's for sure: she was a talented all-around athlete who developed world-class ambitions very early in life.

One day, when she was nine and running errands with her mom, a real estate agent, Heil started flipping through a *Sports Illustrated* she'd noticed on the magazine rack in a store. The issue was all about the Barcelona Olympics, and Heil couldn't put it down. She asked her mom to buy the magazine for her, then pored over it, carefully cutting out pictures of the athletes and taping them to her bedroom walls and into her school binders. From that point forward, whenever anyone asked her what she wanted to be when she grew up, there was no hesitation: "An Olympian." She just didn't know which sport she'd compete in yet. Though she excelled at swimming and swam competitively, her body type wasn't quite right for the sport; really great swimmers tend to be taller. Eventually, seeing freestyle skier Jean-Luc Brassard win gold for Canada at the 1994 Olympics in Lillehammer, Norway, helped her make up her mind. "I can do that," she thought.

Freestyle skiers negotiate moguls while making aerial jumps, but that description doesn't begin to do the sport justice. Part lightning descent of bumpy slopes, part gravity-defying Cirque du Soleil–style acrobatics (and for dual moguls, part mesmerizing balletic synchronicity), freestyle skiing requires you to be both a daredevil and a control freak. "It's like this really weird dichotomy," Heil explains. "A combination of pure guts and letting it go at the same time."

Her father had taught her to ski moguls before she even knew it was a sport, and she'd done her first jump at the age of seven, on a little hill in Edmonton that took her "probably six seconds" to get down. Freestyle mogul skiing, then, felt completely natural. Joyful too.

Her parents encouraged her when she said she wanted to get serious about competition, with the proviso that school had to remain a priority. At fourteen, she marched in to see her school principal and politely informed him of her plans. "I told him that I wanted to compete in the Salt Lake City Olympics in four years, and that to do that, I was going to need a vigorous training schedule and [had to] travel to competitions." To accomplish her goal, she concluded, she'd require some flexibility. The principal jumped on board—something about Heil's manner must have made it clear that this train wasn't stopping—and arranged for her to take some of her courses online.

The combination of ambition, steely determination, and tremendous natural ability took Heil very far, very fast: by sixteen, she was winning national titles, and the following year, she won three World Cup medals. By grade twelve, she was completely accustomed to negotiating with teachers about homework and due dates; to join the World Cup circuit, she'd had to miss three months of school, so she did her coursework online while touring the world and maintaining a punishing training and competition schedule.

Although Heil loved what she was doing, there were days—after a bad competition, say, or a crash on the slopes—when she wanted to quit. The issue wasn't that she was fed up with the rigour or yearned for the life of a normal teenager. (Staying focused has never been difficult, she says with a smile.) The problem was that she was in almost constant pain. Her body needed time to recover, but Heil was too driven to sit around nursing an injury. And by the time she was seventeen, she had plenty of them: terrible shin splints were a daily reality, as was excruciating lower back pain. Moguls are hard on the body, especially the back and knees.

Throw in the hard landings of the jumps, and her body was taking a real beating. There were other injuries too: concussions, torn tendons, a broken thumb . . .

Chronic pain didn't keep her off the hill, even when she had to hobble to her ski boots to get them on. But it did drain the fun out of skiing. She kept going because as bad as her injuries were, she didn't think they were doing any lasting damage; they were soft-tissue injuries, for the most part. It wasn't like she'd broken a femur, for heaven's sake. She could ski through the pain. That's what Olympians do. She was certainly not giving up her dream of standing on an Olympic podium. One day she came across a quotation: "Be comfortable with being uncomfortable." She posted it on the back of her bedroom door.

Still, there was no getting around the fact that she couldn't train the way her rivals did. Freestyle mogul skiing is a power sport—a thirty-second burst that requires every ounce of energy a skier has. Flying down a steep slope of deeply grooved moguls, the skier hits a jump a third of the way down, lands on more moguls, and takes another jump farther down the hill. Most skiers train in the gym with box jumps—leaping from a standstill onto a wooden box at least two feet high—and other big-muscle power drills, like squats. But that was out for the question for Heil. Her body just couldn't handle it.

She continued to excel anyway, because her technical ability was extraordinary. Freestyle skiing, she explains, is "about touch, about having a fast-twitch muscle, and [about] feeling the snow"—all of which she had in spades. Her age, speed, and technique ensured she got a lot of attention, and if press accounts from the time are anything to go by, she wowed the ski world in 2001. But even Jennifer Heil was surprised when she qualified for the Canadian Olympic Team. Despite what she'd so confidently told her high school principal when she was fourteen, she hadn't really counted on making the team for Salt Lake City. The 2006 games, yes, but 2002? It felt almost too good to be true.

Expectations of her were high. As one Canadian sports reporter put it a few months before the games began in Salt Lake City, "Last year she was the ski sensation of the nation. This year we find out if she's the real deal." Talk about pressure.

Heil was tremendously excited, but she was also overwhelmed. She'd worked so hard to get this far and wanted to do well, but she couldn't shake the feeling that she just wasn't ready. And the atmosphere in the athletes' village was surreal. Once, NHL superstar Mario Lemieux plopped down beside her and said hello. She froze. Why would someone like him even notice someone like her? And then it hit her: they were teammates. "Oh, right," she realized. "I'm supposed to be here." But prime time felt very different than anything she'd experienced before. She was accustomed to flying down a hill past a handful of spectators, most of them supportive parents like her own who'd come to see their kids compete; in Salt Lake City, roaring crowds lined the slopes.

When her turn came, she skied fast and executed two perfect jumps, which put her in second place. Then a skier on the Japanese team beat her out by that one one-hundredth of a point and bumped her to third place. Meanwhile, the best skier in the world, Kari Traa of Norway, who'd come first in the qualifying round for the event, was at the top of the mountain, about to fly out of the starting gate. Concern was written all over the face of Heil's coach, but she herself hadn't yet realized what he'd already figured out: barring a disaster for Traa, a three-time Olympian and Heil's idol, Heil would wind up in fourth place. And that's exactly what happened.

At first, fourth place seemed pretty awesome. Because freestyle moguls is a judged event—time matters, but so does form, and judging can be subjective—there was some grumbling by other athletes and those in the media that Heil had deserved more points, and that her performance should have put her on the podium. But she wasn't the one complaining. "Knowing I had done my best and couldn't have done more, I was happy with that," she recalled.

A few days later, though, away from the pageantry and hype, she didn't feel quite so happy anymore. Reflecting on her performance, she had to admit that in fact, she hadn't done all she could. Her jumps simply hadn't measured up. It was a tough truth to face. As she told a reporter, "At the end of the day I realized I was so not prepared. It was shocking. I had never been shown how to do a squat in the gym . . . I was relying 100 percent on my skiing ability and didn't have the athletic ability, hadn't developed into an athlete. It was not a great feeling knowing I could have done more, wishing that I had done more."

Heil decided she had to do something about the pain

she was in so she could train the way she needed to. She consulted with her coach, Dominic Gauthier, and at his urging made the shocking decision to take a year off from competition and spend it in the gym repairing her body and building her strength so she could return stronger and better the following season.

It was a gamble—if she dropped out, she might lose her place on the World Cup team. Those negotiation skills she'd honed in high school proved helpful again: invoking the team's injury clause, she arranged to take time off, but even so there was a lot of pushback. Some members of the organization warned her that she might have jeopardized her spot, injury clause or no injury clause. She was invited to the summer training camp, and there, teammates also urged her not to rush into a decision. There was clucking in the media; she'd come so close in Salt Lake, so why stop now? "There was a lot of attention," she says ruefully—and not the kind she was used to receiving. Nevertheless, she refused to capitulate to her doubts or anyone else's. Although she was just eighteen years old, she knew her own mind. This was absolutely the right decision. She was leaving.

But she wasn't going to take it easy. She was going to build her own kind of team.

One of the most important prerequisites for an elite athlete's success is mental toughness. In fact, many coaches and sports psychologists consider it more important than athletic ability. But what is it, exactly? And can it be learned? An in-depth study conducted by British researchers in

2002 surveyed ten superstar athletes—seven men and three women who'd competed internationally at the Olympics or Commonwealth Games—to try to reach a consensus. Half of the athletes were still competing, and half had retired; their sports included swimming, sprinting, and gymnastics. Their task? Define mental toughness.

The participants were asked to describe the qualities they thought the ideal mentally tough performer possessed. They could base their assessment on an idealized image of themselves or on other athletes they knew, but they had to consider *all* aspects of an athlete's life—not just the intense moment of competition. Through group debates and one-on-one interviews, they hammered out a definition: mental toughness consists of a "natural or developed psychological edge" that enables you to cope better than your opponents with the many demands—competition, training, and lifestyle-related demands—sports place on athletes. That "edge" helps you consistently outdistance your opponents by remaining "determined, focused, confident and in control under pressure." (Interestingly the "edge" is also highly individual. Years later, when Heil underwent bio and neurofeedback tests designed to measure her physical and mental responses to stress, she learned that for her, the challenge was learning to turn *off* her focus in between races and training so that her body had sufficient time to recover from the adrenaline overload of competition.)

Next, the athletes broke the idea of mental toughness down into a dozen key components and ranked them in order of importance. You might think that handling the pressure-cooker of competition would be right at the top of the list, but that was number nine. The most important element, they agreed, is having the belief that you can achieve your goal no matter what. Social scientists call this "self-efficacy." The rest of us call it having faith in ourselves, and it has nothing to do with athletic ability and everything to do with confidence.

It was this quality that allowed Jennifer Heil to turn her back on the skiing establishment and take a year off. If she'd doubted her own ability to achieve her goal—which was to get on that podium at the next Olympics—she would've been more likely to listen to the naysayers and succumb to the pressure to keep competing. Staying on the circuit would have been the more comfortable choice, no matter how much physical pain she was in. She would've been able to keep an eye on her competitors and keep pushing herself to ski better. Taking herself out of the game and insisting on doing it her way—not the way ordained by national sports organizations—was gutsy, to say the least. And it seems to have been driven by a rock-solid belief that she could get where she wanted to go if she was honest about what had happened: she'd failed to medal, she said, not because others were better but because she hadn't done her very best. And to do her best, she had to look more closely at what *she* needed in a training regimen.

How do you get that kind of confidence? The answer can be found in the second component on the superstar athletes' list: the ability to bounce back, because you are more determined than anyone else. Mentally tough people *choose* to bounce back, often using failure as motivation to become stronger and perform better, just as Heil did after placing fourth in Salt Lake City (though her "failure" looked like a pretty big success to the rest of the world).

Mentally tough athletes have faith in themselves even though they don't always win. Their self-efficacy is based on caring deeply about winning, but also on knowing they can cope with losing. Competition gives athletes endless opportunities to deal with disappointment and failure, and the mentally tough ones have learned to roll with the punches and bounce right back, ready for another round. Nothing instills confidence quite like going head to head with defeat.

Other attributes on the athletes' list are the ability to motivate yourself, to stay focused despite distractions, to regain control in the face of unexpected stressful events, to push past your physical and emotional limits, and to remain calm in the face of your rivals' performance. Finally, the athletes said mental toughness includes the ability to stay on-task despite any emotional upheaval in their personal lives. And as Heil discovered, the ability to switch off is critical too, so that you can recharge and unwind.

What's fascinating about this list is that ten people with a slew of medals among them decided that the two most important determinants of mental toughness—believing you can achieve a goal and having the ability to pick yourself up after you fail—have nothing whatsoever to do with performance and everything to do with persevering through doubt, disappointment, and other forms of discomfort.

High on the list of attributes of mental toughness is an unshakeable belief that you possess unique qualities that set you apart from your opponents, whether it's your training regimen or characteristics that make you the "right person for the job." Heil had this too. She took a year off from competition, but she didn't take a year off training. She'd just decided she was going to start doing it in a whole new way. Like Stanley McChrystal, who threw the rulebook out the window when he got to Iraq, she'd concluded that the traditional approach didn't work for her. She needed her own program, tailored just for her and run by a team of hand-picked experts. It was an approach that had never been tried before.

The team she and her coach, Dominic Gauthier, assembled included a nutritionist, an osteopath, a sports psychologist, a strength-and-conditioning coach, a physiotherapist, and an athletic therapist who worked with her daily. Rule number one: there were no rules. They weren't going to worry about how other freestyle skiers trained; they were going to focus instead on how to help *Jennifer Heil* become the best freestyle moguls skier she could be. Abandoning conventions—including the idea that a skier should spend as much time on the slopes as possible—freed them up to design a program uniquely tailored to Heil's needs. In 2002, that approach bordered on heretical. Professional sport had an almost military culture; there was a standard training model, and athletes were supposed to fall in line and not question it. Similarly, businesses often create training or human resource programs that are one size fits all—neglecting the fact that every employee has unique needs and will respond in his or her own way.

The philosophy behind Heil's training program was

completely different: she wasn't just an athlete but a first-year student at Montreal's McGill University who needed a program that allowed her to be eighteen years old. She had to spend only a few hours a day in the gym—five or six fewer than most athletes at her level—but she used those hours in a smarter, more targeted fashion, going all the way back to square one. Her athletic therapist, Scott Livingston, later told a reporter that Heil was in so much pain when she began training that she had to "reprogram herself, even in things as basic as walking."

In a national organization for elite athletes, where the focus is on producing winners, it's unlikely she would've been allowed to relearn the fundamentals. There's too much pressure, too little time, and too much competition for scarce resources. But Jenn Heil's team was all about Jenn Heil, and for her, that was life changing. Later, she said that while many people thought she would lose a lot by taking a year off, she "ended up getting more than [she] could ever imagine, mentally and physically." She got to stay in school and control her schedule—that alone was incredibly liberating. She learned about the importance of rest, and about creating balance in her life. Having lived a very cloistered existence within the sporting world for so many years, she found that meeting people outside that world gave her a new sense of perspective.

To buck the establishment takes courage at any age. To do so at eighteen, when a lot of powerful people are shaking their heads and telling you you're making the biggest mistake of your life, is truly remarkable. The path Heil chose was more uncomfortable, at least initially. She was venturing into uncharted territory, without knowing where she'd wind up. And her trainers wouldn't even allow her to pick up a weight, which killed her because she knew her competitors were lifting them every day. Too bad, she was told. She wasn't touching a weight until she'd repaired some of the damage she'd done to her body. Her postural alignment was out of kilter because she'd been compensating for her pain, so she needed to improve her balance, for starters. Otherwise, she might just reinjure herself in whole new ways. It was nerve-racking at first, thinking about how many steps back she was going to have to take before she could start moving forward again. Rewiring all of her movements was a long, painstaking process. It was as if, like Jim Moss, she was learning to walk all over again.

After a few months Heil felt herself getting stronger, and as the winter progressed, the team got her out on the slopes again. Surprise! She was skiing faster and jumping higher, and she felt she had better control than ever before. And for the first time in memory, she began to ski—and live—pain-free.

But there was a problem: Heil was running out of cash. With this athlete-centred way of training came a whole new level of expense. If she was going to continue training outside the auspices of a national organization, the bill was going to come to close to $80,000 a year. Fortunately, Gauthier was friends with J.D. Miller, a Montreal mergers and acquisitions specialist who is also an avid sports fan and long-time supporter of athletes. He'd helped Gauthier cover some of his training costs in the past, and now he stepped in to help Heil, drumming up tens of thousands of dollars from his business connections in Montreal and Edmonton, where Heil was a hometown hero. Miller even invited her to live with his family while she was training.

In 2003, funding skiers' training via private donations was just not done in Canada. Crowdfunding hadn't been invented yet, and athletes typically left it to their coaches and local clubs to scare up the money they needed to compete. But Heil was getting accustomed to being an outlier. Remarkably, it wasn't putting a chip on her shoulder or making her arrogant or strident. Doing things her own way was just helping her get where she wanted to be. She was continuing to improve.

Then the International Olympic Committee (IOC) tossed a major curveball by deciding to throw out the rulebook, or at least one page of it. Prior to 2003, skiers who did flips in competition were automatically disqualified. But that year, the IOC announced that freestyle skiers could make inverted jumps—the kind where the skier's head winds up below her feet. Suddenly, the expectation was that you couldn't win without a flip. Heil, who'd been practising regular jumps her entire career, had never done a flip in her life.

The team went to work, and so did she. She started by learning to do flips on a trampoline, including two that would become her specialties: the 360, which is a flip with one full rotation, and the technically difficult backflip iron cross, where she had to remain suspended upside down in mid-air with her skis crossed for a fraction of a second. Once she'd got the hang of those on a trampoline, she graduated to skiing down a plastic-lined wooden ramp into a swimming pool and doing a flip en route. The hard part was gaining the confidence to try in the first place, but once she found her groove she had a blast, and finally she was ready to try doing a flip on snow. In a matter of months, she'd mastered a whole new skill—during her first year of university, no less, and far from home.

This, it seems, was the real proving ground: striking out on her own to train, live, and ski in an entirely new way. She'd passed with flying colours.

By the winter of 2003, strong and pain-free, Heil had left McGill and returned to competition, and over the next three years she won dozens of medals, including two consecutive golds in the World Cup. When she landed a spot on the 2006 Canadian Olympic Team headed for Turin, Italy, she wasn't surprised. She was ready. Heading into Salt Lake, she'd been skiing on "guts and enthusiasm." Now she had "all the details down." This time, she knew she really couldn't have done anything more to prepare.

She finished first in the qualifying round, with a flawless run that reduced television commentators to blathering excitedly about her "speed and absorption. She's like a fine automobile on a bumpy road!" Heil herself was calm, because she remembered that when Kari Traa had finished first in qualifying in Salt Lake City, she won gold in the final. Between the two rounds, Heil chilled, riding a spin bike in her room, then getting her mom to bring over a turkey and ham sandwich. "I was relaxed for the first time in a long time," she said afterward. "So I knew it was going to be a good day for me."

As she stood at the mountaintop before her final run, she was "ready to own those thirty seconds." Bursting from the ramp in Team Canada's red, white, and black in the glow of a full moon, "she had a beautiful final run down the dimpled mogul course, her legs acting like pogo sticks as she bounced over the bumps, her body defying gravity as she flew into the air for her first two jumps," the *New York Times* reported. First a perfect 360, and then an iron cross backflip. Upside down, she saw the crowd but stayed focused, completing the flip then tearing to the finish line. As she crossed it, she pumped her fist in the air—she knew she'd killed it. The board confirmed it: 26.69 seconds. Gold—Canada's first ever in women's moguls, and the first Canadian medal of the 2006 Olympics. Flashing back to her nine-year-old self gazing at the pictures of Olympians in that *Sports Illustrated*, Heil felt her eyes fill with tears.

She knew she won the medal in no small part because she'd listened to her own instincts and taken control of her training regimen. After the victory celebrations, the commerce student and team player in her began to wonder how she could help other athletes benefit from all she'd learned. Over dinner with Gauthier (now not just her coach but her boyfriend) and J.D. Miller, she hashed out a plan to devise and fund a program tailored to the specific needs of other Canadian Olympic athletes.

They called the program B2ten—the name sprang from the idea of taking a business-like approach to training athletes for the 2010 Olympics—and launched it in Montreal with two employees, Heil and Gauthier. B2ten is now in its tenth year of operation, and Heil, one of the most decorated athletes in her sport, considers it her proudest achievement: the company has raised tens of millions of dollars from donors interested in helping Olympic hopefuls.

Some of the athletes who have benefited from B2ten's support have said that they view the organization as their "fairy godmother." They don't just receive funding and targeted athletic training. They also get guidance on everything from preparing their taxes to dealing with pushy parents and overly aggressive friends and fans (start by disabling Facebook and Twitter during competition) to coping with the pain of losing—and dashing a nation's expectations. Figure skater Patrick Chan said he was so "bummed" after placing fifth at the Vancouver Olympics that he couldn't bear to show his face and stayed in his room, missing out on the entire patriotic experience. He didn't even eat with his teammates. "I made a lot of trips to McDonald's," he said. At B2ten, he could share his feelings with peers who knew exactly what he was going through. When there are only a handful of individuals in the world who can truly understand how you feel, their support is worth its weight in gold.

Sometimes, B2ten calls on experts, occasionally going far afield to find the skills needed. Swimmer Brent Hayden had had great success swimming the 100-metre freestyle, but he was relatively slow off the starting blocks and in the turn. The B2ten team installed race-quality starting blocks at his home swimming pool, brought in South African swimmer Roland Schoeman to work with him, and flew Hayden and his coach to Estonia to meet with a world-class biomechanics expert who made recommendations for tweaks based on sophisticated software he'd developed for training athletes. Hayden went from being one of the slowest starters in the pool to one of the fastest, winning bronze at the 2012 Olympic Games in London. Sometimes the support athletes require isn't training at all. In one instance, the commute between home, work, and a practice facility was gobbling up too much of an athlete's time. B2ten's solution? Kick in money for a used car.

Co-founding the business while still in active competition was yet another revolutionary act for Heil. B2ten never would have happened if she weren't a boundary pusher, and for amateur athletes, it's been a game changer. One athlete whose training was largely funded by the organization recently told the *National Post* that while the "national sports organizations in Canada try very hard to deliver the services the athletes need . . . the way our sport system is structured, it's not really conducive to pushing the envelope, taking big risks and providing an individualized approach to athletes."

B2ten, interestingly enough, wasn't a hard sell to captains of industry, as J.D. Miller found when he went knocking on doors looking for donors. Some with last names like Bronfman and Desmarais were all too happy to help, and they weren't looking for money, publicity, or corporate endorsements in return. The only benefit they got from their investment was the satisfaction of knowing they'd helped Olympians succeed. Many of them had wanted to help nurture champions all along, but they weren't so keen on governmental bureaucracy. B2ten's more efficient, innovative, and businesslike approach made perfect sense to them.

• • • •

In 2007, Heil won four straight World Cup events as well as the coveted Crystal Globe, awarded to the overall World Cup champion for freestyle skiing. And then she decided to sit out the 2008 season. Over the next year, she tended to her knees, resumed individualized training, went back to McGill to work on her degree (world-class success can turn a four-year degree into a much longer slog), and focused on B2ten.

After a successful 2009 season, she headed to the 2010 Olympics in Vancouver, where the "Own the Podium" campaign was in full swing. She'd trained alongside freestyle skier Alexandre Bilodeau, and their workouts were so brutal, they'd set up a "puke bucket" in the corner of the gym and proudly wrote their names on it each time they vomited.

"Nobody gets a medal in the mail," Heil has said. "You have to fight for it." And that means fighting discomfort. "When you have to be in the swimming pool at seven in the morning, having woken up early enough to have your breakfast and have a good workout [first], it's definitely tough. Thank God for coffee."

There were many moments of frustration during training, and managing them was particularly difficult for her, Heil has said, because she's an emotional person. But she's mentally tough too. "You can throw your ski poles around, but that's not going to help," she said. "It's a very tough sport to master," because ten milliseconds matters, and a millimetre can change everything. "But that's also why I love it. I love to see how I can challenge myself. I try to bring that to the table every day."

Heil, a bona fide celebrity by 2010, was favoured to win gold in women's moguls in Vancouver, and after her stellar performance on the freestyle ski circuit, it seemed that she would. But in front of a multitude of cheering fans and millions of Canadians watching at home, she lost to her rival, US skier Hannah Kearney, and ended up with silver. The loss initially felt devastating—not just because Heil was the favourite and had lost on home turf, but because she and the other B2ten Olympic hopefuls were being closely watched. (The organization supported eighteen athletes at the Vancouver games, twelve of whom won medals.)

Nevertheless, her career continued to soar, and a short time later, she won the World Championship. Then, just before the 2011 race season was to begin, she over-extended her landing in a backflip and suffered bone bruises in her knees. Because she'd landed on the backs of her skis rather than centred on both feet, her femurs smashed into her patellas and bruised both kneecaps on the inside. Heil had toughed it out through shin splints and other injuries, but this was a whole new world of pain, the kind that would have sidelined most skiers for more than a year. Yet when her doctors told her she wouldn't suffer any lasting damage if she stayed on the slopes, she decided, like Ray Zahab, to soldier on.

Heil has learned to think about mental fitness in the same way she thinks about physical fitness. In a video for her speaker's bureau, she explains that she trained on a weekly basis with a sports psychologist. She'd visualize herself in challenging scenarios—on a ski hill, in front of a camera, writing a final exam—then try to create the nervous feelings she usually got in those circumstances. "I'd see myself overcoming them and shifting into how I wanted to act. How I wanted to feel calm. How I wanted to feel positive and how I believed in myself."

Nerves, she says, are a "completely normal" part of competition, both on the ski slopes and in the boardroom. No matter how bad the pain or anxiety gets, you must never lose sight of the reason you're putting yourself through all the hardship in the first place. The most important thing is to be "playful, to have that joy" (a word she uses a lot), and "to be willing to explore new things." As long as you can hang on to that, "you'll be able to overcome really difficult challenges."

Heil had always had "razor-sharp focus." But then her close friend, French freestyle skier Sandra Laoura, wound up paralyzed from the waist down after attempting a flip in training. Laoura, who'd won bronze in Turin, had landed on her back in the snow, breaking two vertebrae despite wearing a back protector. She would never walk again.

Heil no longer felt the same about her sport. In January 2011, prior to competing at the World Championships, she announced that she'd be retiring at the end of the season. It was not a decision she took lightly. Nevertheless, she felt the time was right—she'd achieved all her objectives and wanted "to find success off my skis." And then, just four days before the Worlds, she fell during a race in Calgary. She'd fallen only a few times in her career, and after Laoura's accident and the rocky year Heil had had, it was "pretty crushing in the moment."

But Heil still had one goal as a skier: to win World Championship gold in single moguls. (She was a three-time winner for dual moguls, where two skiers tear down the slope at the same time, but she'd never won in single moguls.) She pulled herself together, won the one medal that had eluded her (and took the gold in dual moguls as well), and then hung up her skis. She'd racked up fifty-eight World Cup medals, including twenty-five golds, and was a five-time winner of the coveted Crystal Globe. Oh, and there were those two Olympic medals as well.

Retirement has been a huge adjustment, to say the least. Sport is interesting, she says, because it has a "due date on it . . . a time that you have to move on." But even though she was ready, she admits it's "a very unique experience" to have spent virtually your whole life working toward achieving "finely tuned goals"—and then have to start over. Talking about it, she switches away from first person, as though even now, several years later, she's not really able to own it fully. "Before you were outside, travelling the world, chasing your dream. Your body changes, your community changes almost from one day to the next. In sport, everything is extreme: the payoff, the emotions. It's all very black and white. You have to recalibrate completely."

Her biggest challenge in the first year of retirement—"and this isn't a joke"—was learning how to sit still at university. "I didn't know how to sit all day, and it was excruciating . . . I couldn't focus, and I couldn't sit there. I was shifting all around. I went from being on top of the world's most beautiful mountaintops to a basement with no windows, sitting in a chair." She's still not great at it, she says, but she has learned to take tea breaks and "walk it out."

Why study commerce, of all things? She sees "vast" parallels between the world of sport and the world of business: both are competitive, move at a rapid pace, and require you to adapt. And she believes that all the lessons she learned in sport—perseverance, resilience, commitment, and dedication—are directly transferable. "I see Jenn as someone who has the potential not only to win gold medals," says J.D. Miller, "but to be the CEO of one of Canada's top-twenty companies." After all, from the age of eighteen, she's been conducting herself like a CEO, setting out a bold vision and then finding a dedicated team to help her execute it.

Heil's comfort with discomfort is what helped her reach her goals. She figured out first-hand that doing things differently was risky but could have a big payoff. She's even comfortable with her small tastes of failure, recognizing that they made her stronger, both physically and mentally. Today, she's grateful she came in fourth in Salt Lake. If she hadn't, she might not have taken the time to reset or had the courage to insist on doing things her own way.

Now thirty-three, she lives in Vancouver with Gauthier and their two young children. These days, Heil doesn't define success as winning. In fact, the silver medal she won in Vancouver is in some ways more precious to her than her gold, and she says that so sincerely you have to believe her. The pressure to win had been so intense, and she's proud of how she kept her focus—then managed through her disappointment. Success, she's decided, is less about winning and more about being able to bounce back from a failure—number two on that list of attributes of mental toughness.

But Jenn Heil didn't learn that by reading a study in a peer-reviewed journal. She learned it in part from Gauthier, who was ranked second in the world in men's moguls skiing and then blew out his knee two weeks ahead of his own Olympic shot in 1998 (he wound up placing nineteenth). And she learned it from her own experience. Setbacks and disappointments are all part of sport, she says. "You can fall off a curb and roll your ankle. You can go to the [athletes'] village and get sick. There are so many things you can't control. Sport can be cruel. Clouds might roll in so you have flat light, while your competitor had sunshine. There are endless variables. All you can control is the effort you put in and how you prepare."

It takes a specific kind of courage—and mental toughness—to reinvent the way something is done, and put your own individual needs front and centre. But the payoff is worth the risk. Because even if your first shot wasn't the right one, what you learn about yourself will have made it all worthwhile. And when you're through, you'll be strong enough to know that you can cope with discomfort—but you'll also be ready to succeed.

CHAPTER EIGHT

The Discomfort of Ambiguity

In business, setbacks are inevitable. For start-ups, they are defining. How founders respond to them not only shapes outcomes but influences the whole culture of a new company. At Workbrain, a workplace management software start-up that was sold in 2007 for C$227 million, the tone was set by founder David Ossip's response to crisis. When something terrible happened, Ossip would calmly say, "Unlucky, unlucky," and let the words hang in the air for a few moments before clapping his hands and adding briskly, "Okay, what are we doing next?"

"What's next?" is a question that inevitably follows change—and for most of us, it's one of the scariest questions of all. We shrink from the unknown and the ambiguous. We prefer the certain, the well-trodden, the understood. But nothing new has ever come from that terrain. And when a crisis strikes, it can be a perfect opportunity to embrace the ambiguous.

Ossip's calm, philosophical approach resonated with Daniel Debow, an early principal at Workbrain, not least because it echoed the relentless optimism he'd learned from his maternal grandparents. Holocaust survivors who'd lost their whole families in the war and then started over in Venezuela, only to be uprooted by a revolution in that country, they came to Canada as refugees when Debow's mother was fourteen. Misfortune had proved to them that it was possible to survive anything. Debow's grandfather, who died when Daniel was in elementary school, remains a towering presence in his life. "He had this saying that I tell people all the time: 'If you go to the front door and you knock on the door and they don't let you in, go to the side door and knock. And if they don't let you in at the side door, go to the back door and knock on the back door. And if they don't let you in at the back door, break a window and just get yourself in if you have to get in! Do whatever it takes.'"

Getting Workbrain up and running involved breaking a lot of windows, and bearing a lot of uncertainty about what was next. It's common now for sophisticated algorithms to calculate the optimal hours for shift workers, but it was a new concept in 1999 when Workbrain launched its software, which allowed companies to clock and schedule employees' hours and automate some human resources functions. For the first eighteen months, the team burned through cash without any revenue, then they decided to focus on selling to airlines, where scheduling pilots, flight attendants, maintenance workers, and other personnel is highly complex. Finally, they landed a big customer: British Airways. Champagne all round, and Ossip headed off to an airline show in the States to announce the relationship. The date? September 11, 2001. "Not the best time to launch an airline-focused software company. Half our pipeline disappeared," Debow recalls.

Unlucky, unlucky. And what's next? Ossip calmly observed that events are random, which sometimes works in your favour but sometimes doesn't. As Debow's grandfather might have said with a shrug, "Deal with it."

By 2003, quite a few random good things had happened, and Workbrain's client roster had mushroomed to include Target, Walmart, Best Buy, Citigroup, and American Airlines. To service those clients, the company had hired hundreds of people in Toronto. The weakness of the loonie was helping keep Workbrain afloat: clients were billed in US dollars, but employees and overhead were paid in Canadian dollars, which meant the company could capitalize on the favourable exchange rate. But then came a snag. In the winter of 2003, Ossip went down to the Toys"R"Us headquarters in Wayne, New Jersey, and was greeted by a sign on the door: if you've been in Toronto in the past two weeks, don't come in the building. The SARS crisis had hit Toronto hard, and with it came widespread fear that Torontonians were walking disease vectors. That was on a Thursday. On Friday Ossip flew home, and Saturday morning the executives huddled. CFO Matt Chapman did the math and announced that if American clients wouldn't let Workbrain employees on-site, they'd be out of business in two months. Unlucky, unlucky.

On Sunday, plan B was already in full swing: Workbrain was moving to the United States en masse, relocating to a

Melrose Place–like apartment complex in Atlanta, where the company already had an office. There were phone calls: "Kiss your kids goodbye. We're moving tomorrow." On Monday morning, about seventy-five Workbrain staff flew to Atlanta to keep the US operation going and stayed there for a few months while Debow and Chapman held down the fort in Toronto. That rational reaction—be upset for a minute, then solve the problem—was what Debow saw over and over from Ossip, and it made an indelible impression. Ossip was the big brother of one of his closest friends, so Debow was accustomed to looking up to him. But he'd really had no idea just how *calm* the guy could be under pressure.

Later that same year, the company ran into a backlog problem: it was taking six months to fix software bugs, and customers were threatening to sue. "Fix it," Ossip told Debow, who promptly freaked out. He was the sales and marketing guy—he didn't know the first thing about software or development! Nor did he relish the thought of having to placate clients' chief information officers. "Getting the CIO of Target on the phone is not a fun experience. These people will rip you apart," he explained. Ossip was unmoved. "Figure it out," he said.

To quell his anxiety, Debow did the only thing he knew how to do: ask questions, read books, and talk to experts. He found that if he was willing to push through the scariness of not knowing, the answer would eventually come to him. You just had to understand that at first you were going to feel sick to your stomach. If you could tough it out, though, you would get an answer, and often that answer would be more innovative than the ones offered by people who started with some actual knowledge of the subject. "Beginner's brain," Debow called it, and he found it worked for him, though it sure could be uncomfortable. Like Linda Hasenfratz, who believes you don't have to be 100 percent ready before taking on a new job, Debow thinks you just have to throw yourself into the "not-knowing," even though "at first it feels like you're drowning."

Everyone knows that expertise takes practice—at least ten thousand hours of deliberate, focused practice, according to Florida State University psychologist Anders Ericsson, whose seminal research on the subject was popularized in Malcolm Gladwell's bestselling book *Outliers*. But what if there was a shortcut, some kind of a hack that meant you didn't have to put in all those hours to transform yourself from a rookie to an expert?

Chris Berka believes there is. A statuesque fifty-nine-year-old neuroscientist with a mane of thick golden hair, she is the founder and CEO of Advanced Brain Monitoring (ABM), a Southern California–based business that develops smart, wearable technology with the aim of taking neuroscience out of the lab and into the real world, where its findings can be used to improve people's lives. A few years ago, ABM received financing from the US Defense Advanced Research Projects Agency (DARPA) to explore technologies that would accelerate the pace of learning by a factor of at least two, and Berka and her team began to look closely at shortcuts to expertise.

When a novice begins learning a new task, the way that Debow had to at Workbrain, many different parts of the brain light up, Berka explains, because neurons are firing and trying to make connections. The learning brain, as captured by an EEG, looks a bit like a miniature Christmas tree covered in blinking lights. Because the brain doesn't yet know which connections are the right ones, a bunch of neurons randomly fire at once. But as a skill is practised, the brain prunes away connections that aren't necessary for the task. Eventually, only the regions of the brain that are essential to the task will light up; the more the task is practised, the more quickly and precisely they light up. This pruning of neural energy allows our brains to streamline and strengthen the specific pathways that will be useful to us when performing a particular task.

For their work with DARPA, Berka and her team decided to create a study involving archers (or marksmen, as they are also known). Previous research at the University of Maryland had already mapped what an expert marksman's brain looked like while he was performing his task. Now Berka's team was going to map the brain and heart activity of marksmanship coaches from the Marine Corps in order to find out what was going on, neurologically and physiologically, in the moments leading up to a shot.

Using a cap that fits over the entire skull, with multiple sensors relaying signals to an EEG, the researchers measured the experts' brain wave activity during those moments. Specifically, they wanted to know which of four types of waves were particularly active. Beta waves are the fast ones, and they correlate with high engagement in a task and a high level of alertness. Alpha waves are slightly slower and indicate a state of non-arousal; learning, memory, and concentration are heightened. Theta waves, even slower, occur when we daydream or ruminate—they're the waves that let us drive home without quite remembering how we got there, but they can also induce the most creative state, when our minds flow freely without self-judgment. Delta waves, the slowest of all, occur in the deepest, dreamless sleep. In the lead-up to a shot, it turned out, alpha and theta waves were strongly present in the brains of expert marksmen, and that combination, plus a slowed heart rate, added up to something that Berka describes as "relaxed focus" and the expert marksmen described as being "in the zone" or "one with the target." It turned out that they didn't even need to be anywhere near a gun to trigger that state either—it happened if they were just *imagining* shooting.

Now that the brain's "expert state in action" had been mapped, could it be used as a sort of cheat sheet for novices? If they could make the correct parts of their brain light up, would their archery skills also leap forward? To find out, the researchers gathered three hundred people—some Marines, some laypeople, but no expert archers—for two days of training.

With each novice, Berka's team first looked at the areas of the brain where the alpha and theta waves (both good) and beta and delta waves (in this case, not so welcome) would appear. They also measured heart rate and breathing, which tipped them off to something important: the most anxious subjects, the ones whose hearts were hammering away, were virtually unable to learn. The hormones released by anxiety, Berka explains, reduce our ability to create and retain memories. "We knew after the first couple of subjects that people with high anxiety levels just didn't learn. It was almost a waste of time," she says.

Few people can tune out the discomfort of anxiety. "As soon as we get anxious or fear failure, we have no chance of getting into the expert state. It's just the wrong direction," Berka says. But her hunch was that if the high-strung novices could short-circuit their anxiety, they would learn faster. So they were given a little device to wear on their collars that buzzed in concert with their heart rate—until their brains approached the "expert state," at which point the buzzing stopped. Just slowing their breathing—and being rewarded by reduced buzzing—moved the anxious novices closer to an expert state: their brains showed an increase of alpha frequency. In just one day, 85 percent of the participants were able to achieve a measure of control over their heart rate and move their brain activity closer to the expert state. In so doing, their accuracy at target practice also shot up significantly. "We were happily surprised by that," Berka says. "We thought we could help people control their brains, but that it could happen in one day was surprising. The neuro-feedback training enhanced their ability to get the perfect shot."

The next day, the novices moved from a bow and arrow to a laser-shot rifle. Now that they knew what the expert state felt like, most of them could return to it without the help of the buzzer. "Though we aren't normally conscious of our heart rate or the state of our brains, we can easily be made conscious of them," explains Berka. "Once there is an association—taking a perfect shot and feeling that state—

you have established, we think, a new neural pathway." While hitting the bull's-eye also provides positive reinforcement, Berka says the good feeling of being in the expert state is really all the inducement most people need. Once that connection was made, participants came up with their own cues to get back to that feeling; they found their own "trigger switch" to return to that state of mind. The experts had described a similar kind of switch, saying they could simply flip it on, take the shot, and then flip it off again.

Did the novices become expert marksmen? No. But they did learn much faster—more than twice as fast, in fact—than they would have otherwise. Berka's hunch turned out to be right: if you can quiet your anxiety and learn to mirror the state of an expert's brain, you can progress much faster. And apparently, the neuro-feedback had lasting effects. A month later, 50 percent of the novices had fully retained their ability to induce the expert brain state.

Several things struck me about this research. First, it's revealing that anxiety really is the enemy of progress and basically shuts the gate on learning. The problem isn't discomfort with challenge and change, but how we respond to and manage—or don't manage—it. If the response is to freak out, discomfort becomes a real obstacle to growth and development; if the response is to calm ourselves—via mindfulness, meditation, tuning down the dial, or focusing on optimism—we have a better shot of figuring out how to get past an obstacle or master a challenge. David Ossip must have understood this intuitively. Certainly he understood that modelling calm equanimity when disaster looms can be the best way to rally the troops and keep them moving forward. Ossip's father was an entrepreneur, so he knew a thing or two about the ups and downs of a start-up, and that likely helped him keep his cool whenever a brush fire broke out.

Second, Berka's research shows that not being an expert is no reason to pass on a challenge. The process of learning a skill may not be as hard as we think. Daniel Debow didn't know a thing about software development, but he was able to figure it out, as Ossip knew he would. If you have the right work ethic and a little luck, jumping—or being pushed—into the deep end can expand your sense of your possibilities and also reinforce mental toughness. Yes, it will always "feel a little like drowning," as Debow puts it, but once you find the right "expert state" of mind, the discomfort is manageable and can be reframed as growing pains or an opportunity. Staying calm is the key.

Of course, Debow had a huge advantage when he found himself flailing in the deep end at Workbrain: it was a familiar feeling. In fact, it was a feeling he'd sought out again and again.

The youngest of three children in an upper-middle-class Jewish family in Toronto, Debow grew up thinking about happiness in a different way than most people do: he knew he was responsible for his own. His father is a psychiatrist, his mother a court interpreter, and their parenting style leaned toward "benign indifference," he says with a smile. So while his friends' parents forked over twenty bucks whenever their kids brought home an A, Debow's dad merely asked, "Do you feel good about it?" If Debow said yes, his dad would reply, "Good, because it's for you. It's not for me." His father, whose psychiatry practice was filled with affluent patients, was also very clear that money does not buy happiness. Debow was raised to believe that most people "optimized for the wrong curve: the wealth curve" rather than "the happiness curve." So he always tried to focus on the latter. Otherwise, what was the point?

Like many future entrepreneurs, he wasn't much of a student. It's not that he lacked an interest in learning, though. As a kid, he used to read all night long. He was so exhausted by morning that his parents finally had to confiscate his flashlight and unscrew the light bulbs in his room to be sure he slept. But school bored him. Contrarian by nature—a "fecal agitator," in his mother's words—he'd developed a "very bad habit" by high school: if he thought an assignment lacked intrinsic value, he refused to do it and took a zero instead. There were a lot of zeroes—so many that his life might have turned out quite differently if he hadn't fallen in with a group of nerds who "stayed home and played [the board game] Risk all night long" instead of partying. To them, failing was not cool. They motivated Debow to push himself.

But when he entered the University of Western Ontario, he still hadn't given much thought to what he was going to do with his life. He'd just kind of assumed he'd become a doctor like his dad. Sitting in physics class in second year, however, he realized he didn't care about velocity and acceleration the way the other students did. He dropped the course, closed the door on med school, and switched his major to psychology. After graduation and a year spent visiting Ivy League colleges as a chapter consultant for his fraternity, he applied to a University of Toronto program that would allow him to earn a law degree and an MBA simultaneously over four years. Although he didn't have a burning desire to become a lawyer, he thought a law degree would open doors to other opportunities.

But U of T rejected him. Debow—who, like Stanley McChrystal at West Point, had been aimlessly coasting on his smarts for years—was now wide awake. He wanted an interview to plead his case. Sorry, that wouldn't be possible. He'd never really dealt with rejection before, and he didn't like the way it felt. He started showing up at the admissions department over the summer, found out which professors sat on the admissions committee, got to know their assistants, hung out where they did, and got friends who were already in the law school to lobby on his behalf. One day on campus he ran into one of the professors on the admissions committee, buttonholed him, and said, "I know you don't do interviews, but I'm here right now. Why don't we chat for a few minutes about why I should be at the law school?" The professor gave him a funny look but agreed. Debow made his pitch. Whatever he said, it was apparently compelling. In late August, he learned he'd been accepted.

Whatever elation he felt about that evaporated entirely on the first day of classes. For the first time in his life, he was out of his league. One-third of his classmates had graduate degrees. "This one went to Yale, that one had a master's from Oxford. I had a three-year psych degree from Western and had earned it mostly by taking multiple-choice exams," he says. He hadn't written essays because his undergraduate courses just weren't designed that way. "I won't say I was traumatized, but I was very mindful of the fact that I had snuck in by the hair of my chinny chin chin."

He spent his entire first year "scared shitless." But instead of letting the fear flatten him, he harnessed its power. For the first time in his life, he studied relentlessly. Not because he cared about getting good grades—that wasn't even on his list—but because he cared about not failing. He put so much pressure on himself that when his grandmother called to wish him luck before his first exam, he broke down and sobbed on the phone. He felt the weight of all his family's struggles and achievements, and he was terrified of letting them down. But he didn't let anyone down. He made the dean's list. It was a turning point in his life.

After second year he landed a job in New York, summering at Sullivan & Cromwell, one of the top corporate law firms in the world. There was a visa snafu, though, so he spent the first month unable to work and instead going to the gym and lounging around. By the time the papers came through, he was bored out of his mind and asked to be put on a file not with a junior lawyer but with a senior person who would work him like a dog. "I want to suffer," he told the person in charge of staffing assignments. "If I don't actually experience this little experiment, I'm going to waste my time here." He got his wish and was assigned to work on a live M&A transaction worth hundreds of millions of dollars—"a small deal" for Sullivan & Cromwell, which may be why the senior associate, who was tied up with a major IPO, let his intern fly to a due diligence meeting in Seattle, telling him not to let it slip that he wasn't a lawyer yet. A junior lawyer from the LA office was also in Seattle, but midway through the second day of meetings, he handed Debow a note that read, "You seem to be doing okay. I have to go now. Good luck." Debow laughs about it now, but he remembers the momentary panic clearly. And ultimately, he says, his "little experiment" paid off. "I actually got a pretty good sense of what it was like to be a real corporate lawyer. And that was great, because I was like, 'This is not what I want to do with my life.'"

Maybe investment banking would be a better fit? The next year he called everyone he knew in Toronto to look for a summer internship, and then they called everyone they knew. Nothing. Then he remembered that he'd met someone from Goldman Sachs when he was backpacking in Europe one summer. At the time, he hadn't even known what Goldman Sachs was, but he'd kept the card. Working at Goldman was the most coveted internship in investment banking, and next to impossible to land. But Debow figured he had nothing to lose and cold-called the guy, who, as fate would have it, turned out to be the best friend of the Sullivan & Cromwell lawyer who'd supervised Debow in New York. And that's how Daniel Debow found himself facing a phalanx of interviewers in Goldman's offices in Toronto, although only Harvard grads were supposed to get those interviews.

It was like a scene in a movie. The first fastball that came hurtling his way was "What the fuck do you think you're doing here? These are *Harvard* interviews. I went to Harvard. Everyone here went to Harvard. What's a University of

Toronto student doing here?" Debow didn't flinch. He was so happy to have made it that far—and so certain that he didn't have a hope in hell of being hired—that he just smiled and told the truth: he'd really wanted an interview and had done everything he could to get one. He found out later that his response had been exactly right. The question was just a test to see how he'd hold up under stress. He got the internship and spent that summer in the London office working on what he says was quite possibly the most interesting deal imaginable (at least in his view at the time)—the world's first cross-border three-way merger between Alcan, Pechiney, and Alusuisse—and discovered something: this was not what he wanted to do with his life either.

Now what? Consulting, he decided as he entered his final year and pocketed a signing bonus from the Boston Consulting Group (BCG), a prestigious global consulting firm. Then in December 1999, David Ossip, the brother of one of his oldest friends, asked if he'd help put together a business plan for a start-up. He was, after all, just months away from finishing his MBA.

Debow leapt at the opportunity to do something interesting on a contract basis. This wasn't some theoretical model or business case like the ones at school; he'd see how to build an honest-to-goodness business from scratch. And he'd do it with friends: his classmate David Stein, who'd also known Ossip forever, joined him, and the two "basically dropped out" to work on the business plan. Right away, Debow knew he'd found his calling. He had so much more autonomy than was possible in a corporate law firm or an investment bank. Not only were the results immediate—

you either made a sale or you didn't—but you weren't constrained by the rules that were in place at advisory firms like BCG. Launching a business was far more alluring than acting as a consultant to one, but . . . he'd didn't have options anymore. He'd taken that signing bonus.

He decided he'd make the most of the Workbrain experience, pitching in wherever he could. His motto became "Don't ask. Find a problem and fix it." Even though much of what he was doing was "decidedly unglamorous"—setting up payroll, recruiting staff—he thrived on constant change and endless problem solving. He also began developing a skill he was keen to learn: sales. "Ironically, they don't teach you how to sell at business school," Debow told me. "There's a marketing department, finance department, operations department—everything—but there's no sales unit. Because selling is a dirty word, especially in Canada."

Partway through the summer, Ossip and his brother, who was also working at the start-up, made Debow an offer: they'd buy out his signing bonus with BCG if he'd come aboard full time. In September, he joined Workbrain. There was all the excitement, panic, and adrenaline of the launch, then the struggle of keeping the business alive during tough times. But by 2003, when Debow had been Workbrain's VP of marketing for three years, he was feeling restless. A mentor convinced him to go back to law school—with a master's he could become a professor; he just needed to find the right program. He eventually found what he was looking for at the Stanford Program in Law, Science & Technology, which brought together his various interests. He still didn't want to practise law, but he liked the idea of acquiring new knowledge that would help him in business.

Not many people would choose to walk away from a company pulling in revenues of $36 million, but Debow had never let money or conventional notions about success drive his decisions. He was still looking to optimize that "happiness curve." Luckily, he had the independence, thanks to money he'd made selling his Workbrain shares, to explore his options.

Ossip was supportive when Debow told him he was going back to school, but everybody else was incredulous. First he'd walked away from coveted jobs with "real firms" to launch some crazy start-up. By some miracle the thing hadn't failed. Now he was leaving again? He needed to have his head examined.

Like Jennifer Heil, Debow knew in his bones that he was doing the right thing. Where others saw danger, he saw opportunity. As a child, he'd had a recurring dream that he was running through library stacks touching all the books on the shelves and magically transferring their contents to his head. At Stanford, he'd actually get to do that—and hang out with some really smart people.

One of those smart people was PayPal co-founder and tech billionaire Peter Thiel, who, it turned out, had also spent a few months at Sullivan & Cromwell in his younger years. When Debow asked why he'd left the firm, Thiel said that he could see the path his life would take if he stayed: first-year associate, fifth-year associate, junior partner, partner. While he'd have a very good life from a financial standpoint—the average partner made $3 to $4 million a year—the prospect of following such a preordained, orderly path was anathema to him. He wanted his life to be a little more random.

Debow identified with Thiel. He too craved unpredictability and wasn't temperamentally well suited for security. "I want to wake up and not know," he explains. "Maybe it's partially because I just like the risk . . . pushing it to the wall . . . seeing what's going to happen. But it's also that when you're uncomfortable, you're at your best. It's when you create things. And it's interesting because when you don't know what's going to happen next, I think it also forces you to rely on yourself—not the firm's brand. The firm is not going to take care of you." It was the message his grandparents had instilled: *Nobody else is going to look out for you. That's why you have to get in the back door or break down the window, or do whatever it takes.*

Midway through Debow's time at Stanford, Ossip decided to take Workbrain public, so Debow pitched in again, flying back and forth to Toronto to help write the company's narrative for the prospectus. Despite a weak market, Workbrain raised close to C$40 million in its offering. After graduating from Stanford, Debow returned to Workbrain to run corporate development. In 2006, Ossip decided to sell the business, and Debow spent that year looking for a buyer. In April 2007, Workbrain sold to Infor Global Solutions, a larger software company, for C$227 million. Debow helped negotiate the deal—and his own exit. Three days later, he boarded a plane for Japan and spent the next four months travelling around Asia with his wife, Jordana Huber. (Debow met the CBC Television reporter in 2002, and they married after he'd returned from Stanford.) In late August, they came home and he said, "Okay. What's next?"

He knew he wasn't going to starve—he'd made some money in the buyout, and he had a JD/MBA and a pretty impressive CV—but he dreaded the idea of a traditional job. Floating around in a backyard pool one day, he and David Stein agreed they could never be working stiffs. "I guess we'll just have to start a company together," Debow said. And that was that. They didn't have a great business idea, or in fact any idea at all beyond knowing that they'd better come up with a damn good idea. They planned to get started immediately. Or tomorrow. Or next week, maybe.

Four months later, Jordana came home from work, found Debow engrossed in a *Call of Duty* video game marathon, and put her foot down. "You need to get out of here," she said, giving him the push he needed, right back into the deep end. He and Stein rented a closet-sized office in downtown Toronto and started going in from nine to four every day, hoping the big idea would materialize. Ossip advised them to keep at it, telling them this was their job now: to go to the office and think. They followed their mentor's instructions, but when you're sitting in a room knocking around ideas and not getting any external validation for your efforts—like money or praise or recognition or even encouragement from peers—it's really hard to press on through the not-knowing.

It's even harder when you slam headlong into skepticism. Debow's long-time mentor, a man he greatly admired, heard about his new business—of sitting around and thinking—and said, "Daniel, you know people don't do that, right? It's really hard to dream something up out of nothing. You're better off investing in four or five existing companies [and becoming a venture capitalist]."

Debow was used to people telling him he was making the wrong choice. He'd even worked as a telemarketer at one point, so he'd become inured to hearing the word no. Ordinarily he just brushed off doubters and used their negativity as motivation: I'll show *them*! But this was no ordinary naysayer. This was a stellar business thinker, his trusted adviser. It was hard not to find his warning unsettling.

Going out socially was not much fun anymore. Many of his friends were now partners in law firms and investment banks, flush with cash and a sense of purpose. Debow came home from every party convinced he'd made a terrible mistake. "What have I done?" he'd ask himself. "I could be doing really well and here I am, sitting in a room with a whiteboard." Every time people asked him what he was up to and he said he was sitting in a room thinking, he got the Look. In Silicon Valley, he was sure, no one would have batted an eye. They *valued* the creative process. In Canada, though, saying "I'm thinking" is the same as announcing, "I'm a loser."

In the past two decades, creativity has moved to the front of the line as *the* skill workers need in a world that's changing at warp speed. (One 2010 survey of fifteen hundred CEOs in thirty-three industries found creativity to be the most crucial contributor to success.) Cognitive psychologist Scott Barry Kaufman, co-author of *Wired to Create*, a 2015 book examining the current research on creativity, has said that the most consistent predictor of creative achievement is "openness to experience"—a mélange of intellectual curiosity, thrill seeking, openness to your emotions, and fantasy. The thing that brings them all together, however, is the drive to explore your inner and outer worlds cognitively and behaviourally: more than anything else, creative people love discovering new knowledge.

While it has long been thought that creative people are emotional right-brain types, recent studies have shown that in fact several parts of the brain are involved in the creative process. According to *Scientific American*, the entire creative process, "from preparation to incubation to illumination to verification," consists of many interacting conscious and unconscious cognitive processes and emotions. Depending on where you're at in the process—and what you're trying to create—different parts of the brain are commandeered to help out.

One of the first researchers to twig to the fact that creativity is a highly complex process with many moving parts was psychologist Frank X. Barron. At the University of California in the 1960s, Barron conducted a groundbreaking study on creativity, inviting leading mathematicians, architects, entrepreneurs, and writers, including Truman Capote and William Carlos Williams, to the Berkeley campus for a few days. There, they completed detailed evaluations and were observed as they mingled. Barron found that while IQ and an academic mindset played a moderately important role in creativity, they didn't fully explain the mysteries of the creative mind. He posited that creativity might be distinct from IQ—a radical notion at the time.

While the Berkeley study revealed that creativity involved a complex stew of elements, Barron found that creative individuals did share certain common traits: they preferred complexity and ambiguity; they had an unusually high tolerance for disarray; they were able to find order in chaos; and they were unconventional, independent individuals who were willing to take risks.

The capacity to tolerate ambiguity is an especially important feature of creativity—you just don't get to have a "Eureka!" moment without wandering in no man's land for an extended period of time. According to Kaufman and his *Wired to Create* co-author, Carolyn Gregoire (and as highly imaginative people well know), it's in those "murky, ambiguous places" that "the creative magic happens." Denise Shekerjian, who interviewed forty recipients of MacArthur Fellowships (the so-called Genius Grants) for her 1990 book, *Uncommon Genius: How Great Ideas Are Born*, reported that creative geniuses had mastered the art of "staying loose": they didn't simply tolerate the uncertainty inherent in the creative process, they embraced it, "often without regard for practicality or efficiency."

Sitting in a room thinking probably doesn't sound very practical or efficient. But it's how Debow and Stein came up with, built, and then sold their next business. Their willingness to think their way through ambiguity paid off, to the tune of C$65 million.

The idea, in the end, was simple: at Workbrain, they'd noticed that millennials frequently stopped by their offices looking for feedback and wondering how they were doing. In early 2008, Debow and Stein started pondering whether employees could get feedback using a social media platform. There was no way to know for sure, so, explains Debow, "You experiment. You're like little scientists. You go off and try something."

The first thing they did was round up some millennials in Toronto for a focus group. During the discussion, some participants became highly emotional—there were tears—about the fact that they weren't receiving any coaching or feedback at work. Some had actually quit their jobs because of it. Debow and Stein knew they were definitely onto something, though both realized it might come to nothing. A lot of ideas that seem amazing at first come to nothing. "Failing won't kill you," says Debow. "The worst that can happen is you'll learn something." In April, they spent a day building a mock-up of a landing page and began showing it around to friends and colleagues. "This wasn't the scientific method," Debow explains. "Basically we wanted to know if the idea was a dumb one or not. Does anybody care about this?" They did, and were also eager to tell their friends about it, so the duo pressed on. They asked a branding person to come up with a name, and from a list of a hundred or so possibilities, they seized on Rypple (because Ripple wasn't available). They liked the idea of feedback having a ripple effect, and soon enough, part of their pitch was that the *y* was representative of questions millennials wanted to ask—and also of Generation Y, who would be the first adopters.

By now their idea was a little more fleshed out: companies would subscribe to their service on a per-user basis, and employees could ask questions about their performance and receive anonymous feedback. For instance, users could build a network of mentors, managers, and peers whose opinions they valued, and after making a presentation, say, ask a question like "How did I do on my presentation today?" Then they'd receive anonymous responses from those in their online network.

In the summer of 2008, armed with a prototype created by one of Debow's fraternity brothers for $5,000 and a 1 percent share in Rypple, Debow and Stein raised more than C$5 million. Lucky, lucky. If they'd waited a few more months, the economic meltdown might have killed their idea before it ever got off the ground. To generate some real-world feedback, they offered the service to a few companies for free, including Capital C, a Toronto-based marketing agency. In December, the *Economist* ran a piece about Rypple that included a rave from Capital C's CEO, Tony Chapman, who said that workers at his company had eagerly embraced the system and he was even using it to solicit feedback from clients.

Lucky, lu—well, maybe not so lucky. "It was too early," says Debow. Suddenly, they were inundated with inquiries, but they didn't even have a finished product yet. It was like "digging for oil and you hit the geyser, but you don't have any way to catch that oil." They still had no idea if the product would work over the long term. They hadn't even figured out what their pricing model would be. Still, David Ossip had taught them well: they had to stay calm and roll it out as fast as they could, to catch some of that oil. But it was frustrating—for them, and for their customers. "We had advertised a Ferrari, and we were delivering beaten-up scooters," says Debow. "So we had to build our customers a Ferrari while they were riding on scooters." But in that period, they did sign some huge clients, including Mozilla, the owner of Firefox. Its human resources chief, Dan Portillo, became such a fan that he quit Mozilla to join Rypple full time.

Lest anyone think the story of Rypple's founding involves a bit of early suffering on the way to easy street, Debow stresses that the "suffering" lasted two years. Their "little idea" was a good start, but they soon discovered it wasn't sustainable. They needed to incorporate new features that would get users more engaged with the system, but when Debow, inspired by a chance meeting with the founder of Foursquare, suggested letting employees award "badges" to each other, he was universally derided. He recalled, however, that the biggest players at Goldman Sachs had proudly displayed the plaques they got each time a deal closed. "These super highly paid people would obsess about these little Lucite deal toys," he explains. "Why? Because it was a way of showing social status or success. You can't bring your Ferrari into the office and let it sit on your desk." So when cajoling, pleading, and arguing didn't work, he insisted that Rypple at least try the badges. Eventually, he was vindicated: those same millennials who didn't feel they were getting enough feedback responded to the status associated with a digital badge. Go figure.

In 2009, when Rypple had "twelve people and some

money in the bank," Debow had a conversation with Lori Goler, Facebook's head of HR (now called the "VP of People"). This was the kind of conversation that can change everything, and Debow had worked every angle to make it happen. Cue triumphant music. Prepare to uncork the champagne.

But Goler said she wanted Rypple to custom-build a system for managing talent. Oh. Debow and Stein were adamant that they weren't going to create unique systems for giant enterprise accounts, as they'd done at Workbrain, because then they'd wind up building a services business, not a software business. Debow took a deep breath. "Lori," he said, "we're going to pass on Facebook as a client. I know you'd pay us an enormous amount of money, but I think it would be better for us to suffer for the next year and figure out what our product really is, and then we'll come to you . . . Otherwise, we'll just build what you think is right and we'll end up building the same thing everybody else did: another HR system for your HR people, and no one else will like it." Debow almost vomited after he hung up the phone. "I was like, 'What the hell did I just do? We tried so hard to get Facebook. They finally said yes! And we said no.'" What's next, indeed?

But it was the right answer. When Rypple went back to Facebook in 2010 with an improved product, they closed a "way better deal," according to Debow. The following year, he and Stein were ready to move on, and they sold Rypple to Salesforce for C$65 million. Today, when he's not teaching at U of T's law school or coaching young entrepreneurs or investing in other companies, Debow is back in his little office. Thinking. He's come up with something—an idea so big that people who hear it tell him it's crazy, so he thinks he's on the right track. After all, most big ideas—Facebook, Airbnb, Uber—sound crazy at first.

Debow frequently speaks at business schools, and when he asks students why they don't want to become entrepreneurs, they tend to go on and on about the risks. He's learned to shut them down quickly. "Okay," he'll say, "let's break down the risks. Are you going to starve? No. Are you going to be put out on the street? No. Worst comes to worst, you'll go live with your parents. Frankly, you're probably a more attractive MBA if you try to build something or work at an early-stage company for a year or two. Your likelihood of getting a job might even increase. So what are you really afraid of? What you are *really* afraid of is that it's uncomfortable to tell people that you're trying something different. It's uncomfortable to tell people you're trying something that could fail. It's uncomfortable that you're doing something that could be a big flop."

It's uncomfortable, in other words, not to know for sure what will happen. Not to know what's next. But as Debow has learned again and again, if you're willing to "literally wash yourself in that pain and discomfort, you might get where you want to go." Even if that's just sitting in a room with a whiteboard, thinking.

CHAPTER NINE

The Importance of Trust

When things are going along fine, maybe even better than fine, making a big change can seem not only pointless but risky. Suppose you deliver mail for Canada Post. You have a secure job, enviable hours, lots of downtime—and you're making more money than you've ever made in your life. Just one problem: your dream is to coach elite basketball players. As you push bills and letters through mail slots, you can't stop thinking how cool it would be if right this minute, you were helping a pro player master a three-point shot.

Or let's say your hoop dreams have come true. You're a centre, a so-called big man in the biggest league of all. At twenty-one, you were drafted by the Cleveland Cavaliers—the fourth pick in the first round, the highest draft pick any Canadian had ever been in the NBA to that point. You're swathed in affirmation and your four-year, $17-million contract catapulted you into the 1 percent. Just one problem: for you, great isn't quite good enough. Your goal is to be even better.

In some ways, making a change when you don't really have to is the hardest kind of change to make. It's easier, and certainly feels less risky, to succumb to the cozy allure of comfort. Things are going so well! But sometimes the biggest risk is not taking one at all. Think of taxi operators pre-Uber or Blockbuster pre-Netflix. Staying the same old course is what got them into trouble.

But how do we resist the gravitational pull of comfort? One way is simply to think about comfort as a trap, not a cushion—as something that may harm you in the long run. To change, you have to turn your back on comfort and actively seek out and embrace discomfort. Doing this is much easier if you have support, someone who walks the path with you, calls out pointers when you need them, and keeps reminding you that this hard thing you're doing is worthwhile. Change, social scientists have discovered, is an extremely long and difficult process, and one that many of us embark on long before we even realize that's what we're doing. That's where the right-hand man comes in: he's there to provide perspective on our goals and the progress we're making toward them. To keep us on track. To be our change agent until we can do it on our own.

Like many kids in Calgary, Dave Love grew up playing hockey. He wasn't a big guy, though, so when he turned twelve and body-checking was allowed, he wisely decided it was time to take up another sport. He tried basketball and instantly fell in love with the game. Compared to other kids his age he was a pretty good shooter, and he liked the freedom of being able to practise on his own, without having to ask his parents for a lift to the rink or lug around a duffel bag full of gear. He also loved the unpredictability of his new sport, marvelling at how fortunes changed in an instant, how the mighty could fall and underdogs could triumph.

When he was thirteen and working as a ball-boy for the Calgary 88's (a minor league team in the now defunct World Basketball League), Love met Chip Engelland, currently the shooting coach for the San Antonio Spurs of the NBA. Engelland, then twenty-eight, took Love under his wing and invited him to practice, giving him a few off-the-cuff pointers. As he did, he taught Love how to shoot and also how to describe shooting. The teenager, who idolized his mentor, took in every word.

He did not, however, morph into Michael Jordan. He was just okay and he knew it, so increasingly he channelled his love of the game into coaching, working at summer camps where he learned how to instruct kids on the nuances of shooting. The idea of making a career of it didn't even cross his mind. Coaching, he says, "was just something I enjoyed doing."

During his second year at Mount Royal College (now Mount Royal University) in Calgary, Love thought about starting his own shooting camp, but he had no idea how to go about it. At twenty, he wasn't used to thinking ahead. Case in point: one day he checked out a school bulletin board, spotted a flyer about a broadcasting program, and thought, "Hey, I haven't picked my major yet, and broadcasting sounds cool. Maybe I should give that a try." After graduating from the program, he got a job in operations at a Calgary TV station. He enjoyed the camaraderie and the schedule: he worked shifts—four days on, four days off—so there was plenty of downtime. Chip Engelland, who'd stayed in touch by phone and had become something between a mentor and an unofficial big brother, suggested he should do something productive with his time off, like starting a shooting camp. Love thought, "Why not? Nobody in town is doing that kind of coaching."

Once Love's shooting camp for teens was up and running, Engelland encouraged him to keep stretching: now he should try to coach athletes at the highest level. In Calgary, that meant the players on the university team, so Love approached the coaches. His pitch was that he'd work for free. Looking back, he thinks that was exactly the right move—not least because many amateur coaches view shooting as a skill they can easily teach themselves. Even in the NBA, shooting coaches are relatively rare; many head coaches consider shooting so fundamental and players so skilled that they believe specialized training isn't required. And yet, the few teams that do have shooting coaches, such as the Spurs and the Dallas Mavericks, have seen dramatic improvements in the accuracy of their players' free throws and three-point shots.

Love, then, was offering the University of Calgary team something that was likely not viewed as an essential service—and something he couldn't claim expertise in. In fact, he was a little worried he'd fall flat on his face, but he reminded himself that at least he'd be falling for free. How mad could anyone get at a volunteer coach? And besides, he did know a thing or two about shooting. He'd definitely proved he could help kids do it, so he knew he could help a bad shooter get better. Still, he understood that he had more to learn than he had to offer, and that to become a good coach, he'd have to take some risks and pay his dues. "I was willing to come in and work and learn and spend my time and teach," he says, "so I think the coach thought, 'There's nothing really that can go wrong here. Let's see how right it [can go].'"

That coach likely didn't understand that Dave Love wasn't just trying to change his players' shots—he was trying to change his own life. His dream of coaching NBA players would have sounded crazy to anyone else, but to Love it was compelling enough that he was happy to spend his days dashing between the TV station, the camps he was running, and practices on campus. As he told the kids he trained, self-discipline and repetition are essential to improvement. As soon as you let yourself get comfortable, you stop getting better.

Tristan Thompson grew up in Brampton, Ontario, a city northwest of Toronto, where his dad drove a truck and his mom drove a bus. The eldest of four brothers in a family that had emigrated from Jamaica, he'd played soccer as a child but, like Dave Love, took up basketball when he was about twelve. Unlike Love, however, he grew several inches between grades seven and nine, and made up his mind to play in the NBA. As a Canadian, Thompson knew it would be an uphill climb, especially since he'd come to the game late: most American kids start playing when they're five years old. But his best shot would be in the United States, where there were far better opportunities for a kid who was serious about making it—and Tristan Thompson, inspired by his dad's work ethic, was very, very serious. He did his research, and before he turned fifteen, he'd left home to play for St. Benedict's Prep School in New Jersey, which is part of the "pipeline to the NBA," according to the *New York Times*.

Even on one of the best high school teams in the United States, Thompson stood out. Tall and lean, he was a force to be reckoned with on the court. He was also bright, had an impressive work ethic and a sunny disposition, and was genuinely well liked—the perfect standard-bearer to carry the flag of Canadian niceness south of the border. But partway through his junior year, in front of thousands of fans, his team endured an embarrassing rout. Thompson got into a mid-game shouting match with his coach, Dan Hurley, and was tossed from the game—and the team. Today player and coach are on good terms, and Hurley, well known for his ability to pick and groom star players, has referred to "that terrible moment" as one he'll always regard as his "biggest regret in coaching—more than any loss." His inability to get that situation under control will "haunt" him, he said, because of the "quality kid" that Thompson was.

Because he'd already committed to the University of Texas and was widely considered one of the top-five high school players in the United States, Thompson could have taken it easy and ridden out his time at St. Benedict's, but instead he transferred to Findlay Prep in Henderson, Nevada, to join an old friend from Toronto, Cory Joseph (now a point guard for the Toronto Raptors). Together, they led Findlay to its first two national championships. Both were named McDonald's All-Americans, joining a list that includes Magic Johnson, Patrick Ewing, Michael Jordan, Shaquille O'Neal, Kobe Bryant, Kevin Durant, and LeBron James. Thompson, then, was on the radar screen of NBA coaches even before he went to Texas, and he played college ball for just one year before the Cavs came calling. It was truly a remarkable achievement, especially for a Canadian who'd come late to the game. In 2016, there were only eleven Canadian players in the NBA.

At six foot nine, Thompson has height, obviously, but number 13 is also a tough, physical defender, unusually quick for a big man, and an outstanding rebounder (in the 2015–16 season, only one other player in the NBA had more offensive rebounds than Thompson did). His knack for grabbing rebounds on the offensive end gives his team a second chance to score after missing a shot—which opposing coaches hate, because the defensive team should be able to "box out" their opponents and control the rebound after a missed shot. If the defender is correctly positioned, the offensive player has to get around him or reach over him—not easy to do without fouling him, but Thompson has both speed and a nose for the ball. Once he grabs the ball, the job is either to make a basket himself or to pass to a teammate who can restart the offence and set up a new play. Converting a second-chance shot is demoralizing for the defence and can change the momentum in a game instantly.

Thompson has a talent for keeping the ball in play and creating second chances.

What Thompson is not, and has never been, is a great shooter. In 2011–12, his first season in the NBA, his field goal percentage (FG%)—baskets actually made versus total attempts—was 43.9 percent. Now, that isn't terrible if you're a shooting threat from all over the court and you take a lot of difficult shots. That same season, DeMarcus Cousins, who'd been the fifth pick in the draft the previous year, shot 44.8 percent playing for the Sacramento Kings. Like Thompson, he's a big man and a centre. But unlike Thompson, who almost never took a shot when he was more than a few feet away from the basket, Cousins attempted shots from all over the court, and landed the occasional three-pointer. So although their FG% was about the same, Cousins averaged 18.1 points per game versus 8.2 for Thompson. And Tristan Thompson's shots were all low-risk—so low-risk that he should have been making a lot more of them than he was.

In the seventies, this would not have mattered all that much. Back then, all baskets in the NBA were worth two points, and the game was all about getting high-percentage shots close to the basket. In that era, Thompson would have fit right in as an old-school centre or power forward: a banger who uses his body to block opponents and clear a path for his teammates, and who's tall enough to dunk (the ultimate high-percentage shot). But basketball has changed a lot over the past several decades. Today, the ability to shoot accurately from all over the court is much more important.

To understand why, look no further than the three-point shot (which is any shot thrown from outside the arc of paint around the basket, roughly twenty-five feet away from the hoop). First introduced in the NBA in the 1979–80 season, more than a decade before Tristan Thompson was born, the three-point shot was initially considered a gimmick by sports fans and reporters—and coaches. Legendary Boston Celtics coach and president Red Auerbach sniffed that it was merely a desperate effort to juice flagging television ratings. But kids across North America—Dave Love included—were mesmerized when Larry Bird lobbed the ball toward the hoop from seemingly impossible distances and managed to score. Back then, Bird was the king of the three-point shot; the Celtics star racked up 649 three-pointers in his thirteen-year career, with a 38 percent accuracy rate. (With Bird on his team, Auerbach probably came around to the idea of the three-pointer pretty quickly.)

Rather than disappearing or lingering as a buzzer-beating Hail Mary play, the three-pointer steadily gained currency. The sheer athleticism required to make the shot (and the new opportunities for shorter players to improve their FG%, because they aren't mobbed by big guys like Thompson far from the basket) gave it dramatic appeal. And then there was the new math: three-pointers add up a whole lot more quickly than two-point shots do.

By the time Tristan Thompson started playing, then, finesse was becoming more important than sheer size, and increasingly, the ability to shoot from the "outside" was considered as important as or more important than having a strong "inside" game. As the action moved farther away from the basket, shooting range increased dramatically. Today, players like the Golden State Warriors' Stephen Curry have made the three-point shot a huge part of the game. Curry long ago surpassed Bird's career record for three-pointers, with 402 in the 2015–16 regular season alone—and he has many more years in the pros ahead of him.

Three-pointers weren't in Tristan Thompson's repertoire, but his two-pointers weren't all that reliable either, and that was starting to hurt him—and his team. In the final quarter of a game, particularly if the score is close, the opposing team intentionally tries to foul weak shooters, counting on them to miss the basket on free throws, thereby giving possession of the ball to their opponents. If Thompson could improve his free throw, he could take the target off his back, stay off the bench late in the game when defence is crucial, and help his team win—bettering his chances of a long-term gig with the Cavs.

One November day in 2012, in his second season with the Cavs, Thompson was goofing around on the court after a morning shoot-around practice when his teammate Jeremy Pargo challenged him to a shooting contest with a difference: both of them had to shoot with their weak hands. Thompson, a leftie, won so effortlessly that Pargo said he should shoot with his right hand all the time. Thompson didn't take the remark seriously. But a week later, he and his coaches decided to try a little test. With a ballboy recording his throws, Thompson took one hundred jump shots with his left hand and one hundred with his right. His right-handed shots were noticeably smoother and more accurate. Wait a sec. Had he been shooting with the wrong hand for the past twelve years?

....

Most scientific research on the subject of change comes from studies on how people kick addictions, but even these extreme problems have the benefit of making the pattern that much clearer: the route to change is both long and predictable, according to the landmark study on the subject. Starting in the late seventies, three psychologists—James O. Prochaska of the University of Rhode Island, John D. Norcross of the University of Scranton, and Carlo C. DiClemente at the University of Maryland—spent twelve years studying how people who set out to change problematic behaviours actually did so, developing what's widely considered the dominant theoretical model. The most striking theme to emerge from their research, which looked at more than one thousand people who'd succeeded in making positive and lasting changes in their lives without the benefit of psychotherapy, is that change doesn't happen overnight. It is a *process*—one with distinct stages—and understanding which stage you're in can make a big difference in terms of managing the discomfort that inevitably accompanies real change.

According to the Stages of Change study, as it is known, the first stage is pre-contemplation. This might just as well be called the stick-your-head-in-the-sand phase because when you're in it, you may not even be aware that you have a problem (although you're usually at least dimly aware that others would say you do). Think of Tristan Thompson, the not-so-stellar shooter. But so what? He's good enough for the NBA, for goodness' sake. Or Dave Love, with his day job and a fun gig coaching kids—nothing wrong with that. Chip Engelland might look at Thompson and say, "He should work on his shot—it's a vulnerability," or look at Love and say, "He's not getting any younger—he should really be chasing his dream." But in the pre-contemplation stage, what others say isn't enough to prompt change. There's no internal motivation or serious intent, which is why people can get stuck there for years.

Only when you've admitted that you have a problem of some sort and you're starting to think about how to act on it do you move on to the contemplation stage. There, you weigh the pros and cons but still don't take action. Like the smoker who vows to kick the habit each morning as he reaches for a cigarette, you can remain in the contemplation phase for a long time. You know there's an issue—a weak shot, a postponed dream—but when you think about the effort you'll have to invest to address it, the mountain feels too steep to climb. Instead, you hang out in the foothills. In this stage, the comfort you derive from your habit—binge eating, for instance, helps you forget how lousy your day was—still seems preferable to the comfort you'll experience when you quit (no more shame about your habit, and you'll look and feel better). This stage can also drag on for a very long time—years, even—because we tend to overestimate the benefit of the status quo, as well as the effort and discomfort required to change.

You'll know you're in the third stage, preparation, when you have both good intentions and an actual plan. You may even have begun taking baby steps toward your goal: studying film of the best shooters in the world, say, or starting up a shooting camp for kids. Even though you're making plans and may have instituted a few behavioural tweaks, however, you still haven't embarked on full-scale change. But you plan to. Soon. This is the decision-making stage, and researchers consider it one of the key phases in determining whether you'll actually be able to stick with whatever change you're preparing to implement. If you've a good plan, you're more likely to be able to execute it.

And this is where Dave Love found himself after working at the TV station for eight years: in the preparation stage. Creating the foundation to achieve his dream. But it probably didn't look like that from the outside. He'd made a career change all right—from broadcasting to mail delivery. One day, a buddy who worked as a letter carrier was talking about what a great job it was, and Love did the math: he was making $35,000 doing shift work at the station while letter carriers earned $50,000, finished at noon, and had weekends off. He'd have his afternoons and weekends free to devote more time to his shooting camps, and he could also volunteer coach at the university weekly instead of bi-weekly. He wanted to coach more but still felt he needed something to fall back on. Never one to put all his eggs in a single basket, he admits, "I like having a safety net for my safety net."

Though Love is pragmatic to the core, his dream was very much alive. Full-time coaching was even beginning to seem like a semi-realistic goal, because the more he coached, the more confidence he had in his abilities. He was even starting to believe he might be able to help a good shooter become great. He'd never know if he didn't get the chance, though, so he decided to try to raise his profile, which is to say, create a profile starting from ground zero.

His first step was to set up an email address—theloveof

thegame@yahoo.com—and start emailing anyone he could think of in Calgary who had any connection whatsoever to basketball. It wasn't an easy process for a shy guy, but he knew that if he was going to get anywhere, he'd have to "throw a million darts and hope one would stick." He went with the pitch that had worked once before: I'll run a shooting clinic for free for your team. Occasionally he felt like an impostor—who was he to think he could do this?—and he never felt comfortable making cold calls, but he wanted this. Really wanted it. So he did what he had to do.

During the off-season in the spring of 2009, he landed his first big gig—as guest coach for the Oklahoma City Thunder for a week—thanks to a recommendation from his old friend Engelland. This was huge for a postal worker from Calgary, but it got even bigger: while in Oklahoma, Love got a call from the Phoenix Suns. Steve Kerr, now head coach of the Golden State Warriors, went to high school with Engelland and had asked for a referral to a shooting coach; Engelland had pointed him to Love, who began packing immediately and spent the summer helping one particular player improve his shot from the free-throw line. He delivered. After a summer with Love, that player's FG% improved by 11 percent. His FG% in the next season was—and remains—his career high. And the gig was important in another way: that summer, Love met Dave Griffin, then the Suns' assistant general manager, for the first time.

After those gigs, and a three-day-a-month contract with the Thunder's farm team, Love thought he was on a roll and other assignments would follow. And yet they didn't. Instead, he went back to delivering the mail and coaching University of Calgary players, and tried to stay on Griffin's radar with occasional phone messages that were rarely returned. A couple of years passed—long years when you've had a taste of NBA action and think you've proven your worth. If this were a Hollywood movie, this is the part where Love would be found in a rundown bar with three-day stubble and a drink in his hand, slurring bitterly that he could've been a contender. This was no movie, though, and Love is an upbeat, can-do guy. He wasn't willing to give up on his dream, so he kept networking and calling his contacts, "not wanting to let it die, but not really having any idea about how to make it grow."

In 2012, by a stroke of luck, Love was invited to be one of the guest coaches at a one-week training camp for the Canadian men's national basketball team. Steve Nash, the celebrated point guard for the Phoenix Suns and the Los Angeles Lakers, and two-time NBA MVP, had just taken over as the team's general manager and was looking to raise its profile and the calibre of its game. A number of guest coaches were brought on board, including Love, who coached forty young players, including some who were already in the NBA, like Tristan Thompson. Because Thompson was there, Dave Griffin, who was now the assistant GM of the Cleveland Cavaliers, also attended.

Love became friendly with the national team's assistant general manager, Rowan Barrett, and they hatched an idea while puzzling over the weakness of some of their players' shooting: maybe Love should coach some players after the tournament. Over the next eight or nine months, the pair brainstormed how it would work. Love pictured being at training camp whenever the team was together, and coaching some of the best players one on one when the team wasn't together. He and Barrett hadn't discussed who was going to foot the bill for all this, but Love's dream was so close to becoming reality that he could almost taste it. Finally, one Friday afternoon in March 2012, when he was out delivering mail, he got the call: the nine-week gig he and Barrett had discussed had been reduced to a three-day thing, and oh, yeah, management preferred that he not speak to the players. Love thought, "Why would I want a job where I'm not allowed to talk?" How could he even coach, then?

That evening he was dejected. But on Saturday, while he was out flying kites with his wife and their two daughters, the general manager of the Charlotte Hornets left him a voicemail: he was looking for a shooting coach for a newly drafted player. Could he call back? Love certainly could. The GM talked about flying him out on Monday for an interview. Love tried not to get excited. He didn't want to be disappointed again, so he focused on the idea that he might get a free flight to the United States.

The following morning, another call came—this time from another area code. Dave Griffin was on the line. Love had once had a "thirty-second conversation" with Griffin when he was with the Suns, but now the Cleveland GM was interested in a much longer talk. He wanted to hire a coach to help Tristan Thompson change his shooting hand; Love had come to mind because like Thompson, he's Canadian. "That's ironic," said Love, "because I just got a call from Charlotte." Griffin promptly booked him a flight to Cleveland and set up an interview. "My head was spinning," Love remembers. He called in sick to work, flew to Ohio, and just forty-eight hours after Griffin's phone call, signed on to work with Tristan Thompson.

Love had decided that until he earned 51 percent of his income from coaching, he wouldn't allow himself to say, "I'm a shooting coach." That day had finally arrived. But he still felt he needed a safety net, so he didn't quit his job outright. Instead he took a leave of absence from Canada Post and gave himself permission to say only, "I'm a letter carrier who happens to have a cool gig with the Cavs."

Whatever the job title, one thing was clear: his dream was coming true.

No NBA player has switched shooting hands midway through his career. Ever. When Jeremy Pargo first floated the notion, Thompson wouldn't entertain it—even after Pargo opined that switching could transform him from a strong player into an all-star. Thompson's first reaction, he later recalled, was to say, "Shut up, man. You don't know what you're talking about. I got to the NBA left-handed. I'm going to stick in the NBA left-handed." (Pargo reportedly retorted, "All right. Well, if you shoot right-handed and you become a better player, just send my check in the mail.") Pargo's comment, though, had jolted Thompson into the contemplation stage, and the more he weighed the pros and cons, the more he came around to the idea that at least trying to change his shooting hand might make sense.

The Cavs were struggling. After LeBron James left for the Miami Heat in 2010, the team fell apart. The Cavs went from an elite franchise to one of the worst, virtually overnight. After winning sixty-one games with James in 2009–10, they lost sixty-three games without him the next season, including a record twenty-six games in a row. Over the next few years, the Cavs got a little better, but no one considered the team one to watch (though Kyrie Irving was a genuine star and among the best point guards in the league). Thompson was a key part of the rebuilding strategy, and he became a steady contributor. But the Cavs front office was always evaluating outside talent and actively looking to improve the team's fortunes via the draft, free agents, and trades. Fresh talent can sideline a player who's good but not outstanding, particularly if he doesn't have marquee value. Thompson's future wasn't at risk, but the more he could help elevate the team, the more secure his position would be. And he had an even more powerful motivator: the desire to become the very best player he could be.

If he could train his right hand to shoot more accurately than his left, he could take his game to a whole new level. "I wanted to take on that challenge. I knew I could keep doing what I'm doing and have a good career and make a lot of money," he told a reporter. "But I don't want to be good. I want to be great."

Thompson also didn't want to become one of those sad old guys talking about what might have been. "A lot of people stick with what they know because they're insecure about putting something new out there and getting embarrassed," he said later. "I don't want to sit here in twelve years and think, 'What if I [had] made that change? Could I have been one of the best power forwards in the league? Could our team have taken a leap?'"

He was inspired by LeBron James, who'd initially faltered after his much-ballyhooed move from the Cavs to the Heat, then worked like a maniac on his body and his game in the off-season and came roaring back with consecutive championships. Thompson knew that what distinguished truly extraordinary players was their refusal to settle. They weren't afraid to, in his words, "put it on the line."

But changing shooting hands wasn't a mere tweak or a tune-up. It was a potentially destabilizing move that could sideline him before he'd really made his mark. Rejecting comfort took enormous courage. Thompson later said that he thought his decision created a stir because "it was, like, 'How can you do all this, have all these accomplishments and a career that's been going uphill . . . Could this make it worse?'" It was a reasonable question.

He had a few things going for him, though: he appeared to be ambidextrous, writing with his left hand but holding his toothbrush with his right (he joked that he was "all messed up" and "still trying to figure [himself] out"). And if things didn't pan out, he could always go back to shooting with his left. But still, he was taking a flyer on Dave Love. Shared citizenship isn't enough to foster the level of trust required to get a player to completely remake his shot.

The big question on everyone's mind, including Thompson's, was whether he was really a leftie or a rightie, or could split the difference. As Love remembers the story, when Thompson first started playing and had demonstrated facility with both hands, he'd asked his coach which hand to shoot with: left, right, or both? His coach asked which hand he wrote with, and when Thompson said, "My left," the matter was decided. He was officially a leftie. In retrospect, says Love, "The better question would have been, 'What hand would you throw a baseball with?'" In Thompson's case, that would have been his right. Still, Love concedes that directing a kid who shows signs of being ambidextrous to shoot with his left hand isn't a bone-headed move. Left-handed shooters are rare in the NBA—in fact, slightly rarer than left-handers in the general population—so there is a perceived advantage because defensive players are more used to blocking on the right. In a high-speed game like basketball, even a millisecond pause as a defensive player readjusts to a leftie can be a benefit. This was Love's big break, and a wildly challenging assignment, but he'd watched enough video of Thompson to be confident that he could help—if Thompson would let him. Love was a little nervous about whether they'd connect on a personal level.

Immediately after the Cavs' season wrapped in the spring of 2013, the pair began working together, first in the Cleveland practice facility and later in Miami, Houston, and Toronto. Chemistry was not a problem. As the months went by, Thompson joked that Love was his summer girlfriend. "Wherever he went, there I was," Love remembers, laughing, "walking ten steps behind." At six feet tall, he looked more like Tristan Thompson's kid brother, but Love went for an avuncular approach. Unlike some of his colleagues' "yelling, demanding" coaching style, his is nurturing. Empathy, in his view, is key to earning a player's trust. "No matter what the change is, or how small *I* may view the change to be," he knows that to the player himself "it probably feels like a rebuild. It's such a personal thing. It's how you've made your livelihood, and it's the identifier to most of the population on who you are as a person." To a player who's been doing things a certain way for years, repositioning his fingers on the ball by half an inch may feel like a seismic shift, and Love believes that barking out orders wouldn't make that shift any easier.

When he first meets a player, Love likes to impart a few of his tried-and-true coaching principles. One of his foundational ones is "There's no escalator to the top." Change takes time and hard work. But Thompson quickly proved that he was up for it. Well aware that shooting wasn't his forte, he was open to changing. That was good news for Love: they were over the first hurdle. His goal wasn't to turn Thompson into the next Stephen Curry, but simply to help him get to the point where his free throws were accurate enough that the other team wouldn't deliberately foul him. Since Thompson had so many other strengths as a player, Love didn't have to "hit a home run," he says. He just had to help Thompson get a little bit better—shooting with a different hand.

The switch started with drills . . . slow, repetitive, mind-numbingly boring drills designed to eliminate every last whisper of superfluous movement from Thompson's shot and force him to focus on the last snap of his wrist before the ball left his hand. The goal was to get the index finger of his right hand to draw an arc straight down from the ceiling to the net. Placing that one finger in precisely the right place is, in Love's view, absolutely critical to getting the ball to leave your hand in exactly the right way. And drilling is the only way to make the movement effortless and automatic, so that a player can deliver in the heat of a game. So Thompson performed each drill, over and over, standing close to the hoop—four hundred snaps of the wrists every workout, two workouts every day.

Since Thompson began playing at the age of eleven, Love calculated, he had probably taken more than a million left-handed shots. Switching hands, then, would be like going back in time to his middle-school self. "You have the experience of an eleven-year-old right now," Love told him. "We're going to work as hard as we can, but no matter how good I am or how hard we work, we can't jam all those years of experience into four months." A change of this magnitude was going to take time.

Not to mention a high threshold for discomfort. So much of basketball is instinctual, but Love was essentially coaching Thompson to abandon the automatic reflexes that had made him a first-round draft pick. He wanted him to be more "un-athletic" during their sessions: to take a split second before shooting and use it to stop, position himself properly, and put the ball in the right place. Like any basketball player, Thompson was accustomed to constant motion, not just up and down the court but side to side as he passed and dribbled. Now his shooting coach wanted him to be as motionless as possible in the split second before initiating his shot so he could be sure his hands were exactly where they needed to be. It was as if the star of an action movie were being asked to hit the pause button for a nanosecond as he leapt from a burning building.

In the Stages of Change model, the fourth stage is action. You're not planning anymore, you're actively modifying your behaviour. You're doing boring drills with your right hand, over and over. Action is a heady stage: when you're in it, it feels as if you already *have* changed—which is why many people in this stage fall into the trap of thinking that they're in the clear and underestimate the effort it will take to maintain the change. Nevertheless, this phase is characterized by wholesale behaviour change. If you were a drinker, you haven't just cut down but have sworn off alcohol altogether. If you were a mailman who wanted to be a coach, you've drafted your letter of resignation to Canada Post. People in the action phase know they're there, and they say things like, "Other people *talk* about changing, but I'm actually doing it." This kind of belief in your own capability, which psychologists call self-efficacy, is important because it reinforces internal motivation and is central to how well you'll be able to maintain the change. Change hasn't been accomplished, researchers say, until you prove that you can maintain it.

The maintenance stage is all about avoiding relapse and consolidating the benefits that started to accrue during the action stage. While maintenance was initially viewed as a static stage—a graduation certificate, of sorts—researchers now see it as a period of continuous change (one that may never really end if your change involved kicking an addiction). It's not just about continuing to shoot with your right hand, but about continuing to improve your FG% too. And for Love, maintenance was about owning his new identity as a shooting coach and removing the safety net provided by his day job.

As the research makes clear, change isn't what occurs at the moment we start exercising, or stop drinking, or leave a relationship, or go back to school, or get a new job. There's a long lead-up to that moment, a process that kicks off well before there are any concrete signs of progress and continues long after it's clear that progress is occurring. It's a very long game, and that's why the ability to withstand discomfort—or better yet, embrace it—is so crucial.

However inconsequential the steps you're taking may seem, they're all part of the process—which is not at all linear. Change is not about steadily climbing a set of stairs but about inching up a slippery spiral—you may lose your footing and slide right back down to a previous stage. Relapse is extremely common for addicts; on average, smokers who've managed to kick the habit tried and failed to quit three times before making it to the maintenance stage. While each tumble back down the spiral *feels* like an abject failure, it isn't necessarily. It may actually constitute another milestone on the long road to your goal.

The same is true for people who are trying to work toward an objective. Relapsing to old habits is something Love has seen many times. Overconfidence, in his view, is the biggest enemy of consistency. "The player thinks, 'My proficiency has got me to the point where I don't need to do these drills or think about these minute details.' They forget that it's *because* they've focused on those minute details that they've gained the proficiency." He wants his players to understand that success can be fleeting and the bottom can fall out at any time. "If you start to believe, 'I'm great,' you're just courting bad luck," he warns them. "You can have three days where you shoot the lights out and think, 'I have this figured out. This is awesome. I can rest on my laurels.' And the next day it's gone."

But partly because they've been working so hard in such a focused way, players may lose their sense of perspective. "You don't jump from being a poor shooter to being a great shooter. You jump from poor to below average," says Love. The goal isn't to become a superstar in a few months, he says. It's to inspire confidence in your team's coach so that when there's a shot you can make, he will expect you to do it rather than screaming frantically at you to pass the ball.

Love's counsel to players who want to make lasting improvements in their game is to stay just uncomfortable enough that they're always on their toes. The players who are least likely to relapse and fall back into bad habits, he says, are those who make a virtue of their ability to tolerate discomfort. They see their willingness to do boring drills till the cows come home as proof of their superiority. "They say, 'Nobody, *nobody* would do this—but I do it. I'm the hardworking, diligent, meticulous one, and that's what makes me great.'"

If you keep at it long enough, those dreary, repetitive drills finally pay off, and players begin to experience the breakthrough moments—the moments when Love feels the joy of coaching. He recalls them with childlike delight, and says his players always know when they're happening because they spot him bouncing up and down on the balls of his feet. "Dave's bouncing again," they crow. A million things can put the spring in his step, he says. "It could be as simple as three good shots in a row on the second workout. We were struggling to get one at a time before. Maybe we are getting somewhere! Just thinking about it makes the hairs on my arm stand up."

The bouncy moment with Thompson came during a pair of exhibition games the Canadian national team played against the Jamaican team in the summer of 2013. This was Thompson's debut as a right-handed player, and it didn't start well. In the first game, he missed both his free throws. But in the second, a close game, he took two shots with seconds to spare and made them both, proving to himself and everyone else that he could shoot well—or well enough—under pressure. He was on his way.

In August, when Thompson played for the Canadian team at the FIBA Americas Championship, he made 86 percent of his free throws—a percentage that would have vaulted him into the top tier of NBA players had he achieved it during the season—and was shooting "extremely well," says Love. "Even too well, in my opinion." He was concerned that Thompson not let success go to his head; lasting success, Love knew, would require him to reject comfort, not embrace it. So the coach gently pointed out that the sample size was small: Thompson had played only about ten games over the course of the tournament.

"As a young, proud, hard-working guy, it's easy to

understand you would want to grab on to any success, and he had a lot of it in that heartbeat of a moment," explains Love. "And it might be really easy to say, 'I've got this figured out.'" Tristan Thompson, though, wanted to see himself as someone who works hard. He didn't want to be that guy who takes the easy way out. Even when he wasn't breaking a sweat, Love says, Thompson always chose to see the work as difficult and to view his discomfort as proof of his diligence. "He's persistent, and even a little stubborn," according to Love. Plus, there was the allure of doing something nobody had ever done before.

In September 2013, Thompson told a reporter that while he wouldn't say that making the transition was easy, he thought it was going "more smoothly than one might assume." Dave Love was encouraged, but he took a longer view: "The real improvement comes in not drifting back to your old habits over the course of the next couple of years."

Just as Judson Brewer does with addicts, Dave Love asks his players to pay close attention to their bad habits—not a huge surprise, really, because in a sense the players he works with have become addicted to shooting a certain way. "Don't focus on performing the movement right," he tells them, "but on catching yourself doing it wrong." It's a fine line, though, because he doesn't want them to overthink things. If they do, they'll freeze. It happens all the time at the free-throw line, to players who've spent thousands of hours practising.

One NBA player Love coached was failing to make half his free throws when they began working together. After a summer of shooting work, he was sinking nine out of ten—but when the new season started, he went right back to his old dismal percentage. Players play on two courts, Love points out: the one they practise on, where perfection matters, and the one in the big arena, where their responses have to be reflexive and so ingrained that they don't choke or get rattled by the spectators. "People think, 'These guys get paid all this money to perform a seemingly simple task, so how come they can't do it?'" Love's answer? NBA players aren't invincible, egomaniacal basket-scoring machines. Behind their swagger and million-dollar smiles, they have the same anxieties that we all do—only they don't have the luxury of stumbling in private. "They have this one thing that they're good at, and they have to do it in front of twenty thousand strangers. When they go out on the court, players are basically wearing their underwear," he says. "These are human beings"—and ones who may have even more difficulty changing because they're pounded by waves of derision each time they make a mistake.

One player Love worked with but won't name—"He's twenty-six years old, he's smart, he's making millions of dollars a year, and he's going bald"—confessed that everyone in junior high had laughed at him when he airballed a shot. One time. The laughter never stopped ringing in his ears. He worried people would ridicule him again—so much so that he couldn't move forward. Change felt too uncomfortable.

Aaron Gordon, a power forward for the Orlando Magic who's been working with Love since he was drafted two years ago, doesn't feel that way at all. "I understand that sometimes it's not going to be all fun, and I'm okay with that," he told me after a practice with Love. He sticks with it, he said, because "I truly want to be great at what I do." He declared that his shot is already "exponentially better," then excitedly began talking about "the catalyst, the one change," that got him where he is now. Well, what was it? "I just moved my thumb in," he said, bursting into laughter. "It sounds simple, right?" After watching Love on the court, coaching players, I know it's not.

A lot of the work happens off the court, Gordon reminded me. He and Love watch films together to "see where my legs are, the degree in my elbow . . . He showed me one video of my free throw where my finger wasn't completely under the ball—I was on a poor shooting streak—and after he showed me that video, I picked it up." Way up: Gordon was the runner-up in the 2016 NBA Slam Dunk Contest.

Part of his progress, Gordon says, is related to feeling that Dave Love is on his side and has his back. Change is easier when you have a right-hand man who's rooting for you to succeed. "He's himself. He's very genuine and transparent. A lot of coaches aren't," Gordon said. "What I mean by 'transparent' is he'll tell you exactly what is on his mind. A lot of coaches operate from their own agenda. Dave does this completely with the intent of making me a great shooter."

His teammate Elfrid Payton also now works with Love, but Gordon said that a lot of pros are still resistant. "They either feel like they already have their shot mastered, or they don't feel like they have the time to do it. Or they just aren't coachable."

NBA players who can tolerate the discomfort of change tend to be rewarded rather handsomely. Tristan Thompson certainly was. He worked with Love until June 2014, and in the 2014–15 season his FG% was 54.7 percent. The next year, it climbed to 58.8 percent—15 percent higher than it had been in his first season. Although still not known for his shooting—his range remains quite close to the basket, and he has never landed a three-pointer in a game—he's become much more accurate from the free-throw line and is an even more effective offensive rebounder. Thompson's gamble on greatness seems to have paid off: two years after he switched to shooting with his right hand, the Cavs signed him to a five-year US$82-million contract. And the year after that, the team won the NBA championship for the first time ever, with Thompson "doing the dirty work" of defence, as he put it, and effectively shutting down two of the most feared sharpshooters in the game, Stephen Curry and Klay Thompson, whenever he guarded them.

Dave Love's gamble also worked out. Possibly because he came to the NBA in a fashion that could only be called atypical, he remains starry-eyed about the privilege of helping top athletes get better at work they too consider play on some level. Both his parents and his sister are teachers, and his dad likes to point out that he's joined the firm now too. Love's subject, of course, is change, but he says there's no final exam. Instead, the process of change is like a road trip that never ends. It's something he learned from his own right-hand man, Chip Engelland, who helped him get started on his own journey.

Have fun along the way, Love tells his players—but

don't stop the drills, because you're never going to arrive at your destination. There will always be room for improvement. "Even Steph Curry, possibly the greatest shooter ever to live, probably still thinks, 'Aw, man, there are a lot of open shots that I missed. I should be making those shots!'" If so, maybe Love could help. Stranger things have already happened.

CHAPTER TEN

Finding the Meaning

"What made you most uncomfortable?" I ask, knowing the question sounds absurd. After all, Maher Arar has endured discomfort of a form most of us can't imagine, and should never have to. Wrongly imprisoned in a foreign country, tortured repeatedly, his name forever tainted with the whiff of wrongdoing, Arar has experienced discomfort at a level that seems intolerable. But the discomfort I'm asking about—and the change that was required—is what followed those events.

The biggest discomfort, he tells me in his gentle way, was learning to accept that none of it could be undone. However unjust it felt, however wrongly accused he was—and was publicly acknowledged to be—nobody and nothing could return to him the years he had lost. "My life was taken in a direction that it was impossible to reverse," he tells me. "There is no way to switch the clock back."

On September 25, 2002, Maher Arar kissed his wife,

Monia Mazigh, and his children goodbye, feeling optimistic. They had spent the summer in his wife's childhood home in Tunis, saving a bit of money by living simply and visiting relatives. But Arar had got word of a possible consulting contract with his former employer, a technology firm called Mathworks, and was heading home early. Because he was in the midst of starting his own tech firm, money was tight—and in the wake of first the dot-com bust and then the economic chill that followed 9/11, business was slow. He promised to call Mazigh as soon as he got to Ottawa, a journey that would take a full day thanks to the circuitous route of the airline ticket he'd bought on points. But Mazigh would not hear from or speak with her husband again for almost a year.

Thanks to a lengthy public inquiry held in Canada in 2006, the facts of Arar's misfortune aren't in dispute—though you wouldn't know that from a glance at the Internet. There, you can still find suggestions of links to terrorism. That none of it is true is just part of the injustice Arar continues to live with. "I was on a path to build a start-up, and all of a sudden my life changed and was taken in a direction it is almost impossible to reverse. Let's face it, my name is still out there," he says now. "My name is tarnished, of course—that too. It's all related." Accepting that there is no turning back the clock has been one of the hardest things he's had to face.

Instead of arriving in Ottawa as planned, Maher Arar found himself detained after landing at New York's John F. Kennedy airport from Zurich on September 26, 2002. Pulled aside at immigration, he was questioned for hours, fingerprinted, and photographed. His queries were ignored, and his request for a lawyer—or even just a phone call—was denied. Instead, he was grilled about his mosque in Ottawa, his views of Osama bin Laden and Iraq, even his relatives.

And he was also questioned about an Ottawa acquaintance, Abdullah Almalki. He knew Almalki, but barely—Arar was colleagues with Almalki's younger brother, Nazih. The men interrogating him—a combination of New York police and FBI agents—pulled out a rental lease agreement he'd signed in 1997, to show him that Almalki was listed as the witness. The reason was innocent—Arar had called his friend Nazih to witness the document, and when he couldn't show, he sent his brother Abdullah—but the fact that the US authorities had the document was chilling. Arar was asked to voluntarily "return" to Syria—the country of his birth, but a place he'd left at age seventeen and had never visited since. He was Canadian, he explained, and had been for two decades. Canada is where he wanted to go. That was where he had studied to be an engineer, where he'd met and married his wife, and where he was raising two young children. Again he asked for a lawyer, and again he was ignored. At one point he was asked to sign an immigration form without being shown its contents. Famished, sleep-deprived, and terrified, he signed it.

Arar was held for a day—without food or any sleep—and then transferred to the Metropolitan Detention Center in Brooklyn, where he would be held until October 7. On October 2, he was finally permitted a telephone call: he called his mother-in-law in Ottawa, telling her he was being detained and feared deportation to Syria.

At last Mazigh knew where her husband was, but that was all she knew. She would learn the extent of his suffering only later. Panicked at the silence, she had already been in touch with Canadian officials in both Tunisia—where she remained, waiting for her young son's Canadian passport—and Ottawa. She had learned nothing of any use. Now she knew that her husband was being held with no explanation, and that he might be deported to Syria. Mazigh knew enough about Syria—a virtual police state with a reputation for torture—to be frantic with worry. She was initially assured by Canadian officials in New York that Arar would not be deported—it made no sense, given his Canadian citizenship. This was one of the oddities of the case. The American government had called Syria a "rogue state" and the State Department had called it a "state sponsor of terrorism." Sending a suspected terrorist to a hotbed of terrorism seemed like an odd move.

While Mazigh waited to find out what would happen next, the *New York Times* wrote a small story about Arar's arrest: it was enough to attract attention from the Canadian media. She agreed to speak to them—and felt a great sense of relief at unburdening herself, even to strangers. And she could see that public attention might keep the Canadian government focused on helping her husband.

On October 8, 2002, Maher Arar was put on a private government plane and flown to Jordan. It appeared that Syria was refusing to allow the plane to land, but Jordan would take him. And so they arrived in Amman at 3:00 a.m., and Arar was forced into the back seat of a car, blindfolded, made to lie down, and beaten any time he spoke. After being held for a day, he was driven across the border to Far' Falastin, a prison run by the Syrian military intelligence.

By the time Arar arrived at the prison, he was exhausted, frustrated, terrified, and utterly demoralized. As he was interrogated—and threatened with veiled references to torture—he decided he would confess to anything. Early in the morning of October 10, Arar was taken to the cell where he would spend the next ten months. It has often been described as "grave-like"—because it was just three feet by six feet by seven feet. It had no natural light, but plenty of rats. The next day he was beaten for the first time, in an interrogation that lasted eighteen hours. When he was asked if he'd attended an al-Qaeda training camp in Afghanistan, he confessed, even though he had never been to Afghanistan. It's impossible for anyone who hasn't undergone torture—both physical and mental—to understand why someone would confess to something untrue. But Arar would have said almost anything at that moment. Unfortunately, this admission would turn out to be the only thing the Syrians had on him—and it would wind up casting a shadow on his reputation at home.

The torture and interrogations continued for months. When—on six separate occasions—the Canadian consul visited him, Arar was too terrified of his captors (who were in the room for each visit) to tell the truth about what was happening. The Canadian officials reported back to Ottawa—and Mazigh—that Arar was doing fine. They had no idea what he was enduring. Mazigh never really believed them, but she had no way of proving her husband was in crisis. By December, he was in mental distress—on three separate occasions his mind became a jumble of terrible memories and he began to scream uncontrollably. As for the happy memories of his previous life, it was as if they'd been wiped from his brain: he couldn't recall them anymore.

In January 2003, Canada's minister of foreign affairs, Bill Graham, spoke to his Syrian counterpart, asking for Arar's release. There was no official reply. Not long after, *Time* magazine published an article called "The Challenge of Terror," which used Arar as an example of how Canada had become a "way-station" for terrorists. There was still no evidence that Arar had done anything wrong.

Canadian officials reassured Mazigh that they were doing all they could to free Arar, but she wanted them to declare his innocence and demand his release. Her lawyer suggested having the RCMP write a letter declaring that her husband had no known link to terrorist activity. But the Mounties merely said they couldn't involve themselves in an ongoing investigation, and said they had played no "role relative to Mr. Arar's present situation."

By then, Mazigh was convinced that Canadian law enforcement had somehow been involved in her husband's arrest, though she couldn't prove it. But a previous late-night visit to their home by the RCMP and a request to meet—with no follow-up—now seemed suspicious. When the Liberal government tried to send a letter to the Syrians urging Arar's release and reassuring them that Canada had no concerns about him, the note was again blocked—internally, by law enforcement.

In July, the *Ottawa Citizen* broke the story that someone inside the RCMP may have given information to American authorities that led to Arar's arrest and subsequent deportation. This revelation came shortly after the Syrian Human Rights Committee confirmed that Arar was being tortured in Syria. These were stunning developments, and terrible in their implications. But at least the media was consumed by the story again.

Back in Syria, suffering the terrible conditions of Falastin prison, Arar received a seventh consular visit in August 2003—after four months with no contact. This time he risked torture by blurting out the dreadful circumstances in which he was living. Although he was terrified of the consequences, it had a positive effect. Within a week, he was transferred to another prison, called Sednaya. There he was placed in a communal cell, which after nine months of solitary confinement "felt like heaven." But he was also shocked to see Abdullah Almalki, who had been arrested on a visit to family in Damascus. Almalki told Arar that he had also been at Falastin, and that he had been tortured there—as he was at Sednaya.

Soon Arar faced new trouble in Syria: the government had decided to try him, not in a military court or even in civilian court, but in the Supreme State Security Court. It was impossible for anyone in Canada—including Mazigh—to know if the process would give the Syrians licence to keep Maher or be used as a means of releasing him. What was known was that the court didn't have a stellar record of recognizing the constitutional rights of the accused.

The answer to came soon enough. About a month after his transfer to Sednaya, on Sunday, October 5, 2003—more than a full year after his initial detainment—Arar was released to Canadian officials, who let him shower at the consul's residence and then promptly put him on a plane back home.

His homecoming was emotional and hard. A permanently changed Arar arrived in Montreal to be greeted by Mazigh, his family, and dozens of reporters and camera crews. As thrilled as she was to have her husband home, Mazigh could see that he wasn't returning as the same person. Instead of the bright-eyed, funny man with the ready smile, this Arar was a sad, frightened man with pain and humiliation lurking just behind faint hope. His fear was evident in some of the first words he spoke: "I'm really scared. Are you sure it's all over and they won't put me in prison again?"

And so began Arar's long path to regaining a feeling of normalcy. In the beginning, the smallest thing could set him off—the sound of a child crying or a bad smell. His anxiety and fear lurked like a beast, waiting to pounce. He had nightmares and felt so mistrustful of the world that the simplest of activities—like paying a bill over the Internet—became things to shrink from. His entire world, even inside his own home, seemed full of danger.

And the emotional discomfort wasn't over now that he was home. Always a deeply private person, Arar found himself the subject of intense negative scrutiny. He was still coping with being accused of things he'd never done. And now that he was safe from his torturers, he realized that his name was still linked with terrorism. It didn't help that some of the very agencies that had wrongly accused him in the first place refused to admit to their mistake and instead leaked false information to the press. His fellow Canadians heard the name Arar and assumed he'd done something suspicious. Who winds up in prison if they're blameless? And so, after his release, he had an incredibly difficult decision to make: he could fight to clear his name—a public fight that would pit him against powerful and secretive law enforcement agencies—or he could retreat into his own life and the privacy he desperately craved. "I had to spend some time before I went public, to think about the implications," he told me—not just for him but also for his family. In the end, however, it wasn't really a choice. "Going public became a necessity for me. There was no other way to fight back."

Part of what drove him was that after his return, Arar continued to deal with accusations—leaked "information" from unnamed sources, saying that he'd made many trips to Afghanistan; that he had never actually been tortured, only "mistreated"; and that he had associated with members of an Ottawa al-Qaeda "cell." Arar and his family became concerned for their personal safety.

One of the hardest moments came a month after his release, when Arar decided to hold a press conference to detail exactly what he had gone through. He and Mazigh and their supporters had also decided that they needed to press the government to find out the truth about his ordeal. "The past year has been a nightmare, and I have spent the past few weeks at home trying to learn how to live with what happened to me. I know that the only way I will ever be able to move on in my life and have a future is if I can find out why this happened to me," he said.

And while that public fight was going on, Arar continued to wrestle with private demons. He suffered post-traumatic stress, including feelings of depression and anxiety that he denied for a long time—not just to doctors, who wanted him to take anti-depressants, but to himself. Symptoms of anxiety, like tightness in his chest or trouble breathing, he put down to stress. It took a long time for him to connect news reports on certain subjects with the strong negative reaction they evoked. When he collapsed in the street in 2014, he was forced to admit that his mental anguish was taking a serious toll. He finally agreed to take medication to help with his anxiety.

Public vindication came for Arar and his family in the form of a lengthy inquiry. Established in February 2004, it was headed by Dennis O'Connor, associate chief justice of Ontario. In his findings, published in 2006, Justice O'Connor found that on every count, Maher Arar had told the truth. And been terribly wronged. "I am able to say categorically that there is no evidence to indicate that Mr. Arar has committed any offence or that his activities constitute a threat to the security of Canada," the judge wrote.

As for the RCMP, it was found to have provided American authorities with false information about Arar, linking him in the most tenuous way to an investigation into terrorist activities in Ottawa. Arar was also the victim of a smear campaign by Canadian government sources after his return to Canada, O'Connor noted. "Canadian officials leaked confidential and sometimes inaccurate information about the case to the media for the purpose of damaging Mr. Arar's reputation or protecting their self-interest or government interests," according to the report.

As a result of the inquiry, the commissioner of the

RCMP, Giuliano Zaccardelli, eventually resigned. The government of Canada issued a formal apology to Arar, and paid him $10.5 million in compensation, plus $1 million to cover his legal fees.

Arar's vindication went even further. In September 2015—almost thirteen years after his torture in that Syrian prison—the RCMP laid criminal torture charges against Colonel George Salloum, who had been a highly ranked Syrian military intelligence officer. It was mostly a symbolic move, but also a crucial one for Arar: the very organization that had been found indirectly responsible for his ordeal was now seeking to arrest one of his torturers. Since many had doubted Arar's account, it added a new degree of credibility to the case.

But even once the inquiry was done and the Canadian government had issued its formal apology, Arar still couldn't find a way to settle his feelings about the whole experience. There was always something nagging at him inside. How could he cope with the injustice of his situation? How could he accept the feeling of loss—of time, of reputation, and of trust? It's the question I asked him too: how does he cope with the injustice? "You're asking for my secrets, right? I'll tell you my secrets. I always try to put myself in the shoes of the people who made the decision to send me to Syria," he explains. "I'm not justifying what they did, but it helps me." He can see how the post-9/11 atmosphere of fear led to terrible missteps. But empathy helps Arar come to terms with those actions. "I still say they were wrong to do it, but they were trying to protect their country. Trying to understand that helps me decrease the anger in my heart, and the bitterness." But he adds, "To say I don't have bitterness would be a lie. I'm not Gandhi." And here—amazingly, really—Arar offers up a small chuckle.

He makes it clear, however, that this exercise in empathy does not extend to the men in Syria who physically tortured him: there is no forgiveness or understanding for them. But there is another secret he uses to cope with his feelings about them. He believes they themselves are victims of their own torture. In other words, to perpetrate that kind of atrocity on another human being is so foreign to nature that those who do it must themselves suffer. "I don't think anyone who tortures another human being can sleep at night. No matter how they try to hide it, they can't have a normal life."

Ten years after his release and seven years after his exoneration, Arar made a conscious decision to disappear. He stopped giving media interviews, stopped appearing in public to speak. He realized that even his most ardent supporters saw him only as a victim. And that is not how he wanted to see himself. He was fed up with that label. It wasn't the word itself that bugged him—after all, he certainly was victimized—but that to be a victim is an all-consuming thing. You are only that, defined by that. And he knew he was more than that. So who was he?

Before his life was thrown off-track, he was a self-described workaholic who would routinely put in twelve-hour days. He wasn't overly interested in world events or anything much beyond his job and his wife and two children. Arar says his youthful self was self-centred. At the time of his arrest, he was on the cusp of seeking financing for his business, which was a tool to help design and build computer chips. He resolved to turn back the clock in at least this one way. "I'm still young, even if I do have grey hair," he says, smiling. "I went back to doing what I loved."

And so he launched CauseSquare, a mobile app that is designed to modernize how people donate to charity. Fast and easy to use, it offers a one-click donation method, while also using push notices and game-like features to engage users with charities. Although it's still in the early stages, Arar is excited about its potential. His ambition is for it to be big—maybe not Facebook or Amazon big, but a $100-million-dollar-a-year company within five to ten years. The app follows a monthly subscription model, and it's for-profit. He may not be as driven by the almighty buck as he once was, but he thinks a for-profit structure ensures discipline and drive.

Starting the business—and remerging from his self-imposed exile—awakened some of the old demons. Arar realized that his name would forever be associated with his ordeal, and that the facts around that ordeal are not always told accurately. He has learned the hard way that the Internet doesn't fact-check. He can ignore the random ugliness of some social media—he's learned not to read it—but he can't control the fact that when you search his name online, lies and misrepresentations still appear. His strategy, however, is not to dwell on what he can't control. "I have to move on with life, and assume things will work and be fine."

Once he had quieted—not cured, not really ever—his post-traumatic stress, Arar still had to tackle the remaining discomfort: the sheer injustice of what happened to him. There is, he realized early, no way to undo the wrong that was done. So how to tolerate that? It's a form of discomfort that he did not want to live with.

For Arar, the answer came in finding a deeper meaning in his ordeal. Specifically, he came to realize that because of what had happened to him, important changes occurred. Among them is the fact that after he went public with his case, the *Washington Post* broke the story about rendition—a little-known government program that could be used to deport people without due process, even to countries known to practise torture. "Had I stayed silent, no one would know what was going on in secret. My case helped other people," Arar says, and that brings him comfort. And, he says, it keeps him from self-pity or wondering, "Why me?"

In fact, when he considers the positive impact his story has had, he believes it may all have been worth it. He was one of the reasons that the overzealous actions of law enforcement officials were exposed, and as a result, ordinary citizens took up the call for transparency and due process. By assigning meaning to his ordeal, Arar is able to come to peace with it.

"One person suffered, yes. And my family too. But at the end of the day, the pendulum has swung back in favour of civil rights. If this hadn't happened, maybe the world would be worse today."

When I set out to write a book about change, I was fuelled by a desire to understand why we humans find change so hard. Having written a book about the importance of asking the right questions, I couldn't help wondering why, even once we've mastered that skill, we find it so hard to listen to the answers. Why, even when we know we should make a change, we still don't make it happen. It felt important to understand because more than ever, we are going to need to know how to change.

Change is about adapting to new circumstances, or adapting our thinking about old ones. And in this complex world—in business and as individuals—the need to adapt and evolve isn't just a nice skill to have, it's critical to success. But the problem with change is that it's uncomfortable. Sometimes intensely uncomfortable. So how do we learn to accept that discomfort? The real way to do it is to find the meaning that justifies the discomfort, just as Maher Arar did.

The first step may be acknowledging that we are feeling discomfort. Barry Pokroy describes this as making the unconscious conscious. Pokroy is a clinical psychologist with penetrating blue eyes and a broad accent that instantly places him as South African. He has helped individuals change personally in his therapy practice for decades, but more recently he has taken the theories he used there into a corporate setting. He argues that the same methods apply, because change for a large group involves each person changing individually.

Pokroy likes to use as an analogy the children's story about the princess and the frog prince who retrieves her golden ball for her. Most of us will remember the story from childhood, but Pokroy attaches a whole new meaning to it. The princess in the story represents us, he explains, and her beloved golden ball is the status quo, or our comfort zone. When she loses the ball down a deep well, a frog offers to get it back for her, but only if he can live with her in the castle forever after. The frog represents change, Pokroy says, and like the princess in the story, we find change repellent. So desperate is she to get her ball back, however, that the princess grudgingly accepts the change. Really grudgingly! For some time, she is disgusted by the frog even as she fulfils her promise to him. But gradually she accepts him, and one day she embraces him and kisses him affectionately. Out pops a handsome prince. The analogy is obvious—when we overcome our discomfort with (or even aversion to) change, the reward will follow.

Pokroy uses the story because it gives people he's counselling a shorthand. "Stop holding on to your golden ball," is something that might get said at a meeting. Or, "It's time to kiss the frog prince." But his real goal is helping individuals understand that change is a process. We don't change our behaviour by focusing on the behaviour alone. It requires a shift in attitude or mindset. And that forces us to really think about our actions (make them conscious), which in turn will help us understand them better, or give us insight into them. Only then can we change the way we see them.

It's clear from stories like Arar's, and many of the others in this book, that it's easier for us to accept change if we have a narrative for why it's happening. In other words, if we invest the change with meaning. The reason that is so effective is because once we have attached meaning to something, we shift our attitude about it. And behavioural change follows quite naturally.

But making the unconscious conscious isn't easy.

Because we are fearful of uncertainty—our brains dislike ambiguity intensely—we cling to the status quo, or our golden ball. The secret is to be aware of that weakness, observe where it affects us, and then stop.

Like the star athlete who can improve only by keeping his motor skills from becoming automated, we have to find ways to keep our mental processes from becoming routine—at least when we want to change or solve a complex problem. This involves slowing our thinking down—and it's important to accept that at first it may not feel good. Effortful thinking is just that—it requires effort and energy, and since our neurological wiring urges us against it, it can bring great discomfort. But change, like the handsome prince in the children's story, is a powerful reward. And as we learn that letting go of our comfort, and first tolerating and then embracing the discomfort, will bring us great rewards, it becomes easier to do.

The strategies for dealing with discomfort will vary based on the form that discomfort takes. Reframing the feeling of discomfort as a positive signal—like positive training pain—can make it our ally, rather than our enemy. Choosing to shut out the negative signals can help us keep our older, less thoughtful brain out of the equation and give our prefrontal cortex a shot at winning out. Manufacturing a sense of control—even if it's only in our minds—can minimize discomfort dramatically. Reinventing ourselves can feel like the hardest form of change in the world, but it's actually the easiest if we're doing it in response to pain or discomfort.

As we learn to lean into our discomfort, it will gradually turn into comfort. Our zone of discomfort moves, in other words, as our comfort zone expands. And then our challenge is to stay right there, leaning against that new discomfort. The high-achieving changers among us will learn to identify some forms of comfort as a negative thing—and reject them outright.

Turning away from the discomfort of change is tempting, of course, because you're turning away from risk and the possibility of embarrassment and failure. But you're also turning away from the opportunity to stretch yourself, test yourself, and grow. You're turning away from the possibility—the likelihood—that change will be for the better (and even if it isn't, you will almost certainly learn something worthwhile). Every time we back away from discomfort, change gets a little more frightening. The muscles we need to flex start to atrophy.

Being alert to the possibility of change also requires us to be awake to our own lives. To inhabit them mentally in a way that too often we avoid, daydreaming through half our days. But rather than living on autopilot, we can instead choose to engage more fully with every moment, and the potential each one has for us. The great thing about change is that it's a skill, and once learned, we can improve at it. And then we can learn to enjoy it, as we embrace the beauty of discomfort.

And the really good news is that it's never too late. Whether the answer is to reframe the discomfort, or ignore it, or lean into it, or reinvent yourself, or enlist help, or simply dive in at the deep end, there *is* an answer that will take you from "Why change?" to "But how?" Don't stop looking until you find it.

ACKNOWLEDGMENTS

First and foremost, I owe a debt of gratitude to the people who shared their stories with me—not only because of the time they were willing to spend, but also because these stories are by nature personal. I am aware of the trust they have placed in me, and I'm humbled by it. My hope is that the courage and wisdom they showed in learning to change will be an inspiration to others. And so to Ilana Ben-Ari, Ray Zahab, Jim Moss, Linda Hasenfratz, Frank Hasenfratz, Jessica Watkin, Stanley McChrystal, Jennifer Heil, Daniel Debow, Dave Love, and Maher Arar—my profound thanks. You are an incredible group of people. Your stories moved me.

Judson Brewer's work on mindfulness and habit is groundbreaking and was an inspiration to me. Andy Puddicombe is not only a source for this book but also the "voice in my head" as I continue to use his Headspace app to meditate (though not often enough). Chris Berka's work suggests that we may be able to reverse-engineer ideal brain states—something that has the power to transform the way future generations learn and think, and feels like a game changer to me. My thanks to Aaron Gordon for taking the time out of a pro basketball player's schedule to talk to me.

A book is always a collaborative effort, and often involves many people whose names don't wind up in the finished version but whose thoughts and guidance helped shape it materially. In that category belong Kathleen Hays, Claude Legrand, Reza Satchu, Nick Woodman of GoPro, and Chris Fussell of the McChrystal Group. My thanks to Wendy Dennis for superb editing under the gun. Thanks as well to Barry Pokroy, whom I feel extremely lucky to have met when I did (and because of that meeting I was tempted to change the title of this book to *Dolphins Aren't People*). And thanks also Ted Cadsby, author of *Closing the Mind Gap*, who is present in the book in many ways, and whose influence was so much more than what is represented here.

I owe a huge debt to Terry Stuart, who never fails to inspire and energize me, and who is the only friend for whom I need a notepad at lunch (so I don't miss a thing). He will see his fingerprints all over this.

Thanks to Rick Broadhead, who thought there was something worth doing here, and to Brad Wilson at HarperCollins, who believed in the book long before he had reason to and, as always, made it infinitely better. Thanks as well to the whole team at HarperCollins who helped shape and improve this project—from the incomparable Janice Weaver to Rob Firing, who makes sure it gets read.

How to thank Kate Fillion—one of the best writers,

editors, and coaches in Canada? Your help and guidance were essential to the contemplation of this project, and credit for the title is all yours. If I have a superego outside my head, it may be you. Thank you for everything.

To my friend Maya Carvalho and my twin, Adrian Lang, thank you for your support and love—and also to Corinne Pruzanski, who always sharpens my thinking. My whole family was supportive and encouraging of this project, and I can't express enough gratitude to them for that—especially Julian (who never stops changing), Lauren, and Sam, who had to witness the making of this book firsthand. And to Geoff, who is wise and kind and knows not only how to change but also how to bring out the best in others, including (and maybe especially) me—my gratitude and love.

NOTES

Introduction

In 2012 I wrote a book about the importance of asking questions, and how fostering our natural curiosity leads to innovation and change. But in talking to people about that book, *The Power of Why* (Toronto: HarperCollins, 2012), I realized that wanting to change and actually changing—or identifying a problem and acting on the solution—are two very different things. We humans find it very hard to change, even if we have an idea about what that change should be. I wanted to find out why, and how we might overcome that. The stories in this book are intentionally personal. That is because while businesses also need to understand the process of change and the impediments to it, all change is really about individuals. The interviews for this book took place between September 2015 and October 2016.

Uber's story is well known, and I drew on multiple sources here. But it's also worth noting that in a world where "uberization" is happening to many industries, it's also, at the time of writing, happening to Uber itself.

New businesses are emerging to connect drivers and riders directly, cutting out the middleman altogether.

Clayton Christensen's best-known work is *The Innovator's Dilemma: When New Technologies Cause Great Firms to Fail*, a seminal text first published in 1997 (Boston: Harvard Business School Press). But the Harvard professor and businessman has written many other books on the subject of disruption. I have had the pleasure of interviewing him several times.

The idea that creativity relies on breadth of knowledge and experience, not just depth, is from Adam Grant, "How to Raise a Creative Child. Step One: Back Off," *New York Times* (January 30, 2016). The article can be found at http://www.nytimes.com/2016/01/31/opinion/sunday/how-to-raise-a-creative-child-step-one-back-off.html.

Chapter One: The Benefits of Discomfort

It seems that the concept of the helicopter parent was first articulated in *Between Parent and Teenager* by Dr. Haim Ginott (New York: Scribner, 1969). On page 18, Dr. Ginott quotes one teen as saying, "Mother hovers over me like a helicopter." The actual term "helicopter parent" was coined in 1990 by researchers Foster Cline and Jim Fay, and cited in *Parenting with Love and Logic: Teaching Children Responsibility* (Carol Stream, IL: NavPress Publishing, 1990). Julie Lythcott-Haims, the former dean of freshmen at Stanford University, used the term in her book, *How to Raise an Adult: Break Free of the Overparenting Trap and Prepare Your Kid for Success* (New York: St. Martin's Press, 2015). Lythcott-Haims was interviewed by Emma Brown of the *Washington Post* for an article that cites authors of similar books on overparenting, including Jessica Lahey (*The Gift of*

Failure: How the Best Parents Learn to Let Go So Their Children Can Succeed [New York: HarperCollins, 2015]) and Jennifer Senior (*All Joy and No Fun: The Paradox of Modern Parenthood* [New York: HarperCollins, 2014]). That article, from October 16,2015, can be found at http://www.washingtonpost.com/news/education/wp/2015/10/16/former-stanford-dean-explains-why-helicopter-parenting-is-ruining-a-generation-of-children/.

I had a very personal reaction to the decision made by the Ontario Soccer Association to no longer keep score in games for young players. Brendan Kennedy wrote an article about that decision: "Ontario Youth Soccer to Stop Keeping Score, Standings," *Toronto Star* (February 16, 2013), (http://www.thestar.com/sports/soccer/2013/02/16/ontario_youth_soccer_to_stop_keeping_score_standings.html). The idea, according to the article, was to allow each child to get equal attention from the coaches regardless of skill level. "The new plan," wrote Kennedy, "dubbed 'Wellness to World Cup Long-Term Player Development,' establishes a common training curriculum for all age levels and abilities, more training of volunteer coaches, [and] increased ratio of practices to games while putting a greater emphasis on building fundamental skills." Other provinces implemented similar plans.

On April 30, 2012, when the policy was still being developed, I wrote an opinion piece for the *Globe and Mail* (http://www.theglobeandmail.com/opinion/why-we-need-to-teach-our-kids-how-to-fail/article4104820), arguing that "learning to accept a loss gracefully and rallying to try again is arguably more important than taking home a trophy." I received feedback from several coaches who were convinced I had misunderstood the point of the policy, which was about developing all players equally.

Ellen Winner has done a lot of research into the low rate at which gifted children go on to produce truly revolutionary work in their field. She is quoted in Adam Grant's *New York Times* article "How to Raise a Creative Child. Step One: Back Off" (cited in the introduction). More of her thoughts on the subject can be found in "The Origins and Ends of Giftedness," *American Psychologist* 55, no. 1 (January 2000): 159–69, http://psycnet.apa.org/journals/amp/55/1/159. The quotation I cite is from page 165.

Winner's paper attempts to debunk some myths about giftedness, including the view that it is entirely a product of training (Anders Ericsson's ten thousand hours of deliberate practice, popularized by Malcolm Gladwell). Winner argues that there is evidence the brains of gifted children are different and show signs of enhanced right-hemisphere development. But most interesting is her conclusion that few gifted children go on to achieve greatness in their fields (defined by Winner as "big C creative," meaning that they break new ground or innovate in a meaningful way) because the skills required to excel as a professional adult are so different from those inherent in a gifted child.

One recurring theme with gifted children is that the ease with which they accomplish things early in life leaves them with a sense that they haven't earned the reward or praise they get. It's a subject explored at length by psychologist Carol Dweck in her research on "mindsets" (specifically growth vs. fixed mindsets). The idea is that individuals with a growth mindset are more likely to see themselves as able to improve and change, while those with a fixed mindset may not persevere because they believe their skills will stay the same. Because early skill and knowledge acquisition comes easily for gifted children, they are more likely to fall into a fixed mindset.

Trigger warnings have received a lot of attention in the past few years, including in the *Atlantic* magazine article I cited (http://www.theatlantic.com/magazine/archive/2015/09/the-coddling-of-the-american-mind/399356/). But Kate Manne, a philosophy professor at Cornell University, defended their use in an opinion piece published not long after the *Atlantic* article: "Why I Use Trigger Warnings," *New York Times* (September 19, 2015), http://www.nytimes.com/2015/09/20/opinion/sunday/why-i-use-trigger-warnings.html. Her chief argument is that trigger warnings allow her to respect the sensitivities of or even trauma experienced by her students while still exploring—not avoiding—the material. The issue has stirred a great deal of debate, but the concern that students are being shielded from ideas at precisely the time when they should be asked to stretch themselves intellectually resonated with me.

Robert Reich recounted his personal story to me during an interview I conducted for CBC Television in 2015.

I conducted several interviews with Ilana Ben-Ari and also participated in a half-day workshop for educators held at Twenty One Toys in Toronto.

Judson Brewer and I spoke several times, and his work on the intersection of mindfulness and habit (or addiction) was one of the first pieces of research I explored when I began this book. Brewer has also had a book published, *The Craving Mind: From Cigarettes to Smartphones to Love—Why We Get Hooked and How We Can Break Bad Habits* (New Haven: Yale University Press, 2017). The quotes here are based on my conversations with him, but he also kindly allowed me to read his manuscript while I was researching and writing. Those who are interested in finding out more about Brewer's reasoning should watch his 2015 TED Talk, available online at https://www.ted.com/talks/judson_brewer_a_simple_way_to_break_a_bad_habit. He also wrote about the addictive "habit loop" in the August 2014 blog post "Is Mindfulness an Emerging Treatment for Addiction?" available at http://www.rehabs.com/pro-talk-articles/is-mindfulness-an-emerging-treatment-for-addictions/.

Jon Kabat-Zinn is considered the father of modern mindfulness (it's a centuries-old practice) because he popularized it as a tool to relieve stress and other negative emotions. His book *Full Catastrophe Living: Using the Wisdom of Your Body and Mind to Face Stress, Pain, and Illness* (New York: Delacorte, 1990) introduced the concept to the mainstream. Kabat-Zinn also wrote *Wherever You Go, There You Are: Mindfulness Meditation in Everyday Life* (New York: Hyperion, 1994), a useful guide to practising mindfulness. The quote cited is taken from a *New Yorker* profile of Andy Puddicombe, the founder of the Headspace app: Lizzie Widdicombe, "The Higher Life" (July 6 and 13, 2015), http://www.newyorker.com/magazine/2015/07/06/the-higher-life.

I interviewed Andy Puddicombe twice—once on television, and later, when I began researching change, by telephone. I had been using Puddicome's Headspace app for some time when I interviewed him, so our first encounter was slightly disconcerting: up until that moment, his voice had only been "in my head." Other helpful sources included the *New Yorker* profile cited above and Nilufer Atik, "Man Behind Meditation App Goes from Monk to Millionaire," *Telegraph* (London) (October 12, 2014), http://www.telegraph.co.uk/men/the-filter/11154773/Man-behind-meditation-app-goes-from-monk-to-millionaire.html.

I am relying on Puddicombe for information on the number of users of Headspace. As a private company, Headspace does not typically disclose those numbers, but in 2015, when it raised money privately, it warranted that its user base was three million. By November 2016, news reports were suggesting that Headspace's user base was ten million.

The study about mindfulness and anxiety is John J. Miller, Ken Fletcher, and Jon Kabat-Zinn, "Three-Year Follow-up and Clinical Implications of a Mindfulness Meditation-Based Stress Reduction Intervention in the Treatment of Anxiety Disorders," *General Hospital Psychiatry* 17, no. 3 (May 1995): 192–200.

The 2010 Harvard study, which found that we spend much of our time with wandering minds, was Matthew A. Killingsworth and Daniel T. Gilbert, "A Wandering Mind Is an Unhappy Mind," *Science* 330, no. 6006 (November 12, 2010): 932, http://science.sciencemag.org /content/330/6006/932.

In 2011, neuroscientist Sara Lazar and a team at Harvard found that mindfulness meditation can actually change the structure of the brain: eight weeks of mindfulness-based stress reduction was found to increase cortical thickness in the hippocampus, which governs learning and memory, and in areas of the brain that play a role in emotion regulation and self-referential processing. There were also *decreases* in brain cell volume in the amygdala, which is responsible for fear, anxiety, and stress—and these changes matched the participants' self-reports of their stress levels, indicating that meditation not only changes the brain, but also changes our subjective perception and feelings. In fact, a follow-up study by Lazar's team found that changes in areas of the brain linked to mood and arousal were associated with improvements in how participants said they felt—that is, their psychological well-being. So our subjective experience—improved mood and well-being—does indeed seem to be shifted through meditation as well. See Britta K. Hölzel, James Carmody, Mark Vangel, Christina Congleton, Sita M. Yerramsetti, Tim Gard, Sara W. Lazar, "Mindfulness Practice Leads to Increases in Regional Brain Gray Matter Density," *Psychiatry Research: Neuroimaging* 191, no. 1 (January 30, 2011): 36–43.

Chapter Two: Reframing Discomfort

I met Ray Zahab by chance, on a late-night flight from Vancouver to Toronto, and was immediately struck by his positive personality. Though he had been the subject of a highly successful documentary, *Running the Sahara*, I was unaware of that at the time. In subsequent interviews with Zahab, I learned the details of his amazing story. I also relied on his autobiography, *Running for My Life: On the Extreme Road with Adventure Runner Ray Zahab* (London, ON: Insomniac Press, 2007). Zahab also co-authored, with Steve Pitt, the book *Running to Extremes: Ray Zahab's Amazing Ultramarathon Journey* (Toronto: Penguin, 2011), and he is currently at work on a new project. More about Ray can be found on his website, http://rayzahab.com, and information about impossible2Possible can be found at http://www.impossible2possible.com. In December 2015, Zahab received the Meritorious Service Cross, awarded to Canadians who perform exceptional deeds that bring honour to the country.

Research into the origins and neurological expression of pain is a field of growing importance. After the plasticity of the brain became an acknowledged physiological fact, studies of how pain is experienced had new potential. The theory of non-linearity and assigning "meaning" to pain comes from John Heil and Leslie Podlog, "Pain and Performance," in *The Oxford Handbook for Sport and Performance Psychology*, ed. Shane Murphy (Oxford: Oxford University Press, 2012).

The quote about the brain as "a computer rewiring its own software" is from C. Richard Chapman, Robert P. Tuckett, and Chan Woo Song, "Pain and Stress in a Systems Perspective: Reciprocal Neural, Endocrine and Immune Interactions," *Journal of Pain* 9 (2008): 122–45.

The meta-analysis of pain research studies is from Jonas Tesarz, Alexander K. Schuster, Mechthild Hartmann, Andreas Gerhardt, and Wolfgang Eich, "Pain Perception in Athletes Compared to Normally Active Controls: A Systematic Review with Meta-Analysis," *Pain* 153, no. 6 (June 2012): 1253–62.

Jennifer Heil, the subject of chapter 7, also references "good" and "bad" training pain. The ballet dancers who described pain positively (Helen Thomas and Jennifer Tarr, "Dancers' Perceptions of Pain and Injury: Positive and Negative Effects," *Journal of Dance Medicine and Science* 13 [February 2009]: 51–59) also noted that their sense of control over their pain influenced its severity.

Multiple studies over time have explored how a subject's ability to control pain or unpleasant stimuli alters their perception of it. Some studies have focused on the effects of helplessness, such as Matthias J. Müller's "Will It Hurt Less if I Believe I Can Control It? Influence of Actual and Perceived Control on Perceived Pain Intensity in Healthy Male Individuals: A Randomized Controlled Study," *Journal of Behavioral Medicine* 35, no. 5 (October 2012): 529–37. Other studies focus purely on a sense of control, such as "Self-control and Predictability: Their Effects on Reactions to Aversive Stimulation," *Journal of Personality and Social Psychology* 18, no. 2 (May 1971) 157–62, authored by Ervin Staub, Bernard Tursky, and Gary E. Schwartz. But in both cases, control over pain reduces its effect.

The description of Aron Ralston's ordeal is drawn partially from the Heil and Podlog article cited above, as is the detail about female British rowers viewing injury as a desirable aspect of their sport. The original study of those rowers found that for some sports, a culture of pain acceptance can actually lead to a greater risk of injury.

Chapter Three: The Mindset Factor

The idea that we do not have the tools to measure our own happiness (and in fact don't really think much about it until it hits an extreme—either very happy or very unhappy) struck me as compelling. Research into gratitude has led some psychologists to conclude that the practice of listing the things for which we're grateful should be added to clinical interventions. The quote by the British researchers is from Alex M. Wood, John Maltby, Raphael Gillett, P. Alex Linley, and Stephen Joseph, "The Role of Gratitude in the Development of Social Support, Stress, and Depression: Two Longitudinal Studies," *Journal of Research in Personality* 42, no. 4 (August 2008): 854–71.

Hormesis is a biochemical theory claiming that very low doses of something toxic can actually have a beneficial effect on an organism. The idea is that low doses of a stressor of some sort kick-starts the body's repair response systems. One of the leading experts on hormesis is Suresh I. S. Rattan, one of the authors of a 2008 paper called "Hormesis in Aging," with researchers from the Laboratory of Cellular Ageing, Department of Molecular Biology at the University of Aarhus in Denmark. It explored how a variety of negative stressors, such as irradiation, food restriction, high heat, and hypergravity produced anti-aging and life-extending hormetic effects.

Martin Seligman, considered one of the founders of positive psychology, has written several books, including *Learned Optimism: How to Change Your Mind and Your Life* (New York: Knopf, 1991), *Authentic Happiness: Using the New Positive Psychology to Realize Your Potential for Lasting Fulfillment* (New York: Free Press, 2002), and *Flourish: A Visionary New Understanding of Happiness and Well-being* (New York: Free Press, 2011).

In *Learned Optimism*, he has a test you can take to find your own level of optimism (or pessimism, as the case may be). There is a shortened, adapted version of the test available at http://web.stanford.edu/class/msande271/onlinetools/LearnedOpt.html.

The seminal research on adaptation theory as it relates to happiness was by Philip Brickman, Dan Coates, and Ronnie Janoff-Bulman, "Lottery Winners and Accident Victims: Is Happiness Relative?" *Journal of Personality and Social Psychology* 36, no. 8 (August 1978): 917–27. They found that people who experienced a surprisingly good event (winning the lottery) didn't wind up happier than those who suffered a surprisingly bad event (an accident that left them paralyzed).

Some evidence suggests that a positive attitude may actually lengthen your life. One well-known piece of research on gratitude is Deborah D. Danner, David A. Snowdon, and Wallace V. Friesen, "Positive Emotions in Early Life and Longevity: Findings from the Nun Study," *Journal of Personality and Social Psychology* 80, no. 5 (2001): 804–13. In it, researchers reviewed handwritten autobiographies from 180 Catholic nuns, scored them for positive or negative emotions, and then compared them over a period of twenty-two years to the health and longevity of the nuns. It turns out that positive emotions expressed in early life had a strong correlation with longevity. A PDF of the study can be found at https://www.apa.org/pubs/journals/releases/psp805804.pdf.

The study of gratitude in the workplace is Yeri Cho and Nathanael J. Fast, "Power, Defensive Denigration, and the Assuaging Effect of Gratitude Expression," *Journal of Experimental Social Psychology* 48, no. 3 (May 2012): 778–83.

The study of how university students responded when they practised gratitude was "Counting Blessings versus Burdens: An Experimental Investigation of Gratitude and Subjective Well-being in Daily Life," *Journal of Personality and Social Psychology* 84, no. 2 (February 2003): 377–89. Its authors, Robert Emmons and Michael McCullough, wanted to know if they could establish a causal relationship between gratitude and a sense of well-being. Their theory held out. The researchers also studied adults with a chronic disease to test their theory, and it held out there too.

Chapter Four: Rejecting Discomfort

I interviewed Linda Hasenfratz by telephone and in person in late 2015 and early 2016. I also interviewed her father in the spring of 2016. The quotation from an analyst who was nervous about nepotism is from an article published in *Report on Business* magazine (Dawn Calleja, "Linamar's Drive to $10-Billion," [May 27, 2010], http://www.theglobeandmail.com/report-on-business/rob-magazine/linamars-drive-to-10-billion/article1390169). That article also provided the information on the growth in dollar content per vehicle for Linamar between 2001 and 2010.

The quote from analyst David Tyerman can be found in Mary Theresa Bitti, "Linamar Chief Linda Hasenfratz Continues to Prove She Belongs at the Top," *Financial Post* (December 15, 2014), http://business.financialpost.com/entrepreneur/linamar-chief-linda-hasenfratz-continues-to-prove-she-belongs-at-the-top. Tyerman has followed Linamar since the year 2000. He is now a transportation analyst at Cormark Securities.

There are a lot of people researching rumination, but Yale University psychologist Susan Nolen-Hoeksema is one of the leaders in the field, having conducted several studies linking rumination and depression, as well as rumination and social support. Over the course of years Nolen-Hoeksema conducted many studies, several of which are referred to here, including that of Bay Area residents coping with the Loma Prieta Earthquake of 1989, co-authored with Jannay Morrow ("A Prospective Study of Depression and Posttraumatic Stress Symptoms After a Natural Disaster: the 1989 Loma Prieta Earthquake," *Journal of Personality and Social Psychology* 61, no. 1 [July 1991]: 115–21), the survey of 1300 people, and the research on rumination after the loss of a family member. The quote from her is from an article about her work that references some of the research cited here: Bridget Murray Law, "Probing the Depression-Rumination Cycle," *Monitor on Psychology* 36, no. 10 (2005): 38, http://www.apa.org/monitor/nov05/cycle.aspx.

One of her papers, co-authored with Blair Wisco and Sonja Lyubomirsky, explores how rumination leads not to change but to a heightened attention on the symptoms of a problem. It also explores the strategy of redirecting from a painful subject to a neutral one: "Rethinking Rumination," *Perspectives on Psychological Science* 3, no. 5 (2008): 400–24, http://sonjalyubomirsky.com/wp-content/themes/sonjalyubomirsky/papers/NWL2008.pdf.

Chapter Five: The Comfort of Control

My thanks to the folks at the Canadian National Institute for the Blind, who answered the call when I asked to be put in touch with someone who had experienced the massive change of vision loss. They suggested Jessica Watkin, who was extremely generous (as well as articulate and thoughtful) in sharing her story with me. It is impossible, as many researchers note, for someone who has never experienced vision loss to truly appreciate how difficult it is, but Jessica's first-hand account helped my level of understanding enormously.

The statistics on blindness in the United States come from the website of Cornell University's Institute on Employment and Disability (http://www.disabilitystatistics.org). These stats exclude anyone living in an institution, perhaps because that would capture more elderly people with age-related vision impairment and make the numbers jump dramatically. The statistics for Canada, including those on the number of visually impaired people who are unemployed or live below the poverty line, are courtesy of the Canadian National Institute for the Blind (http://www.cnib.ca). One of the facts that has stayed with me is that a huge majority of late-life vision loss could have been prevented with a regular eye exam.

The information on the rates of depression among those with acquired blindness comes from a study of more than ten thousand people who took part in a nationwide survey, the National Health and Nutrition Examination Survey, between 2005 and 2008: Xinzhi Zhang, et al., "Association between Depression and Functional Vision Loss in Persons 20 Years of Age or Older in the United States, NHANES 2005–2008," *JAMA Ophthalmology* 131, no. 5 (2013): 573–81, http://jamanetwork.com/journals/jamaophthalmology/article-abstract/1660943.

The research regarding late-onset blindness and depression, which showed that 90 percent of people experience depression after vision impairment, is from Roy G. Fitzgerald, "Reactions to Blindness," *Archives of General Psychiatry* 22, no. 4 (April 1970): 370–79. The research about reframing vision loss and how reliant that is on individual personality is from Betsy Zaborowski, "Adjustment to Vision Loss and Blindness: A Process of Reframing and Retraining," *Journal of Rational-Emotive and Cognitive-Behavior Therapy* 15, no. 3 (September 1997): 215–21. In Diego De Leo, Portia Hickey, Gaia Meneghel, and Christopher Henry Cantor, "Blindness, Fear of Sight Loss, and Suicide," *Psychosomatics* 40, no. 4 (July 1999): 339–44, the researchers note that having vision suddenly *restored* can be almost as traumatic as losing it. That study can be accessed at https://www.researchgate.net/publication/12896208_Blindness_Fear_of_Sight_Loss_and_Suicide.

Suzanne Thompson is an expert in health psychology, social psychology, and applied psychology, and she has conducted extensive research into how a sense of control can shape our experience of an illness or emotion, as well as pain. The quote cited here is from "Will It Hurt Less If I Can Control It? A Complex Answer to a Simple Question," *Psychological Bulletin* 90, no. 1 (1981): 89–101. (The quote itself is found on page 95.) The paper also cites a great deal of other research and previous studies, including the one of cardiac patients and the nursing home study.

Chapter Six: The Discomfort of Reinvention

I interviewed Stanley McChrystal twice in early 2016, on television for Bloomberg Television Canada and also off-camera, upon the publication of his book *Team of Teams: New Rules of Engagement for a Complex World* (New York: Penguin, 2015). Some of the material in this chapter is drawn from those conversations, and some is from the book. I also relied upon his earlier book, *My Share of the Task: A Memoir* (New York: Penguin, 2013), for some biographical information. Other quotes and facts are sourced from news articles, as shown below.

The reference to the ultra-secretive JSOC is from John Barry, "The Hidden General," *Newsweek* (June 25, 2006), http://www.newsweek.com/hidden-general-111127.

The information on waning public support for the Iraq War comes from a Gallup poll: Joseph Carroll, "Support for Iraq War Remains Divided," March 24, 2005, Gallup News Service, http://www.gallup.com/poll/15367/support-iraq-war-remains-divided.aspx.

McChrystal's description of change leadership appeared in Lillian Cunningham, "Stanley McChrystal on How to Shake Up the Military," *Washington Post* (May 15, 2015), https://www.washingtonpost.com/news/on-leadership/wp/2015/05/15/gen-stanley-mcchrystal-on-shaking-up-the-military.

Dr. Paulus's research can be found in the article "Differential Brain Activation to Angry Faces by Elite Warfighters: Neural Processing Evidence for Enhanced Threat Detection," *PLoS ONE* 5, no. 4 (April 2010), http://dx.doi.org/10.1371/journal.pone.0010096. The second reference to Dr. Paulus comes from Brian Mockenhaupt, "A State of Military Mind," *Pacific Standard* (June 18, 2012), https://psmag.com/a-state-of-military-mind-9f16f4ec885. Some of Dr. Paulus's subsequent research can be found in Douglas C. Johnson, et al., "Modifying Resilience Mechanisms in At-Risk Individuals: A Controlled Study of Mindfulness Training in Marines Preparing for Deployment," *American Journal of Psychiatry* 171, no. 8 (August 2014): 844–53, http://dx.doi.org/10.1176/appi.ajp.2014.13040502.

Amishi Jha and her colleagues set out to understand how to train soldiers not just to stand at attention but to have "a mind at attention." They also noted that our minds tend to wander more under high-stress circumstances. That may be a defence mechanism, not unlike Linda Hasenfratz's practice of "turning down the dial." One reason, it is postulated, is that we are attempting to cope with our emotional reaction to the stress, which leads to rumination, or a wandering mind. Dr. Jha's co-authored article, "Minds 'At Attention': Mindfulness Training Curbs Attentional Lapses in Military Cohorts," *PLoS ONE* 10, no. 2 (February 2015), is available as a PDF at http://www.amishi.com/lab/wp-content/uploads/Jhaetal_2015.pdf.

Other research notes that while mind wandering does appear to have some positive functions (it allows us, for example, to process complex internal information and goals), it has also been linked to lower psychological well-being. One useful study on that subject is David Stawarczyk, Steve Majerus, Martial Van der Linden, and Arnaud D'Argembeau, "Using the Daydreaming Frequency Scale to Investigate the Relationships between Mind-Wandering, Psychological Well-being and Present-Moment Awareness," *Frontiers in Psychology* 3 (September 2012): 363, https://www.ncbi.nlm.nih.gov/pmc/articles/PMC3457083.

The reference to the *New York Times* calling the war a cataclysm is from an article by Dexter Filkins, "Stanley McChrystal's Long War," *New York Times Magazine* (October 2009), http://www.nytimes.com/2009/10/18/magazine/18Afghanistan-t.html.

The reference to the *Esquire* article refers to this one by Tim Heffernan: http://www.esquire.com/news-politics/news/a5922/who-is-stanley-mcchrystal-051909.

Peter Bergen is CNN's security analyst, and the quote attributed to him is sourced from his book, *Manhunt: The Ten-Year Search for Bin Laden from 9/11 to Abbottabad* (New York: Broadway Books, 2012).

In this chapter I also relied on Robert D. Kaplan, "Man Versus Afghanistan," *Atlantic* (April 2010), http://www.theatlantic.com/magazine/archive/2010/04/man-versus-afghanistan/307983/. And Stan McChrystal wrote about his own transition in this April 22, 2014, post on Linked-In: "Career Curveballs: No Longer a Soldier," https://www.linkedin.com/pulse/20140422120137-86145090-career-curveballs-no-longer-a-soldier.

The article that wound up being the undoing of McChrystal was Michael Hastings, "The Runaway General," *Rolling Stone* (June 22, 2010), http://www.rollingstone.com/politics/news/the-runaway-general-20100622. It caused shockwaves when published, but with the passage of time, it seems like a relatively tame commentary. It's possible to imagine that just a few years later, McChrystal might not have faced a firing squad over his overly candid remarks. This article was published before the real free-for-all of public opinion was unleashed by social media. Given McChrystal's success in adapting to an unwieldy enemy, we have to wonder "what might have been" had he kept his position. But you can read the article and judge for yourself.

Chapter Seven: A Majority of One

I interviewed Jennifer Heil several times, and she was infinitely patient with my questions about milliseconds here and there. I also relied on other sources to verify and add detail to her athletic accomplishments, including the wonderful article "Jennifer Heil: A Story of Talent, Determination and Compassion," *Montrealer* (December 1, 2011), http://www.themontrealeronline.com/2011/12/jennifer-heil.

The quote about the expectations Heil faced pre–Salt Lake City can be found in Terry Jones, "Salt Lake and Sugar Plums," *Edmonton Sun* (November 13, 2002), http://www.canoe.ca/2002GamesColumnistsPreGames/jones_nov13-sun.html.

Heil talks about the shock of realizing that she was not prepared in Mark Masters, "20 Questions: Jenn Heil Speaks on the End of an Era," *National Post* (March 12, 2011), http://news.nationalpost.com/sports/20-questions-jenn-heil-speaks-on-the-end-of-an-era.

The study of elite athletes was summarized in Graham Jones, Sheldon Hanton, and Declan Connaughton, "What Is This Thing Called Mental Toughness? An Investigation of Elite Sport Performers," *Journal of Applied Sport Psychology* 14, no. 3 (January 2002): 205–18. There has been much research on the subject of mental toughness, including later studies that showed differences based on age, experience, and gender. The term is often used in place of "resilience," but psychologists do not believe the two words are interchangeable. That's because resilience is seen as a coping strategy in response to a difficulty, while mental toughness is a quality that can develop over time and will vary from one person to the next.

The reference to Heil learning later about her neuro-feedback tests and what they told her about her body and its responses is from Ken MacQueen, "The Bilodeaus: An Inspiration," *Maclean's* (February 23, 2010), http://www.macleans.ca/news/canada/an-inspiration.

The quotes about Heil's rebuilding of her body and movements come from Vicki Hall, "Ten Years Later, Olympic Champion Jenn Heil's Legacy Still Paying Dividends for Canadian Athletes," *National Post* (February

10, 2016), http://news.nationalpost.com/sports/rio-2016/b2ten-helping-canadian-athletes-reach-olympic-dreams. Heil's description of the importance of her year off comes in part from "Small Wonder: Jenn Heil," *Ski Canada Magazine* (December 11, 2005), http://skicanadamag.com/small-wonder-jenn-heil.

The remark about commentators "blathering excitedly" refers to NBC's coverage of the 2006 Olympics.

Some of the descriptions of Heil's performance in Turin are drawn from Juliet Macur, "Bumpy Start for Americans Is Opening for Canada," *New York Times* (February 12, 2006), http://www.nytimes.com/2006/02/12/sports/olympics/bumpy-start-for-americans-is-opening-for-canada.html. That article also notes that when Heil was on the podium to receive her medal, she was incorrectly identified as an American.

Some of the information about B2ten comes from a story the CBC's Adrienne Arsenault did called "Making the Best Better," which aired on *The National* in October 2013. The quotes from B2ten athletes come from this CBC story, including Patrick Chan's. A link to the story can be found on B2ten's website at http://b2ten.com/news. The quote by an athlete in the *National Post* was from retired bobsled pilot Helen Upperton, a B2ten alum. A link to that article can be found here: http://news.nationalpost.com/sports/olympics/b2ten-helping-canadian-athletes-reach-olympic-dreams.

Heil's comments about not getting a gold medal in the mail and about managing her emotions come from an interview she gave related to her association with Marcelle Cosmetics in 2009. It can be found at https://www.youtube.com/watch?v=-5f1PZbiWYg.

The comments Heil made about visualizing herself in different situations come from a 2013 video shot by her speaker agent, Speakers' Spotlight. It can be found at https://www.youtube.com/watch?v=PbfTj_wf0Ag.

Miller's reference to Heil as someone who could not only win medals but also run a major company came from Ken MacQueen, "Canada's Olympians No. 1: Jennifer Heil," *Maclean's* (December 10, 2009), http://www.macleans.ca/general/canadas-olympians-no-1-jennifer-heil-2/.

Chapter Eight: The Discomfort of Ambiguity

The information about Workbrain and Rypple came from interviews with Daniel Debow.

I came across Chris Berka's work because of my interest in neural plasticity, an area that is exploding in new research and findings. All descriptions of Berka's studies came from my interviews with her.

The references to creativity and its breadth are based on the book *Wired to Create: Unraveling the Mysteries of the Creative Mind*, by Scott Barry Kaufman and Carolyn Gregoire (New York: Penguin Random House, 2015). I also relied for source material on Christie Aschwanden, "'Wired to Create' Shows the Science of a Messy Process," *New York Times* (February 8, 2016), http://www.nytimes.com/2016/02/09/science/book-review-wired-to-create-scotty-barry-kaufman-carolyn-gregoire.html.

The rave review of Rypple appeared in "The Rypple Effect," *Economist* (December 30, 2008). It can be found at http://www.economist.com/node/12863565.

Chapter Nine: The Importance of Trust

Anyone who knows me will chuckle at the idea that I wrote a whole chapter about a professional sport, but this isn't really about the game so much as the people in it. I relied on various sources, including press reports, for the facts and stats about the league and its players, but unless my copy editor has a hidden love for basketball, you will have to forgive any ham-fisted references.

Dave Love came to my attention because he was a Canadian shooting coach in the big leagues; I was curious about how he got there. He would be the first to admit that shooting coaches were a rarity when he started out—and for a long time after. But there is no question that for Love, it was a dream job. You can find out more about him and his shooting camps at http://www.theloveofthegame.com.

Shooting coaches are more common today in the NBA, but not that long ago it was considered part of the general coach's job to improve players' shots. The idea of hiring a specialist in this area has taken off in recent years, thanks in part to the dramatic results that coaches like Love have achieved. See Chris Herring, "Actual Job: Teaching NBA Players to Shoot," *Wall Street Journal* (October 24, 2012), http://www.wsj.com/articles/SB10001424052970203897404578076793880988504.

Tristan Thompson talked about the influence his dad's work ethic had on him in an interview on ESPN after a Cavaliers win in the 2015–16 season.

Thompson's choice of school was no accident: New Jersey is now home to a growing number of "feeder" teams for the NBA. See Harvey Araton, "N.B.A. Pipeline Bypassing New York for New Jersey," *New York Times*

(December 18, 2012), http://www.nytimes.com/2012/12/19/sports/basketball/new-york-basketball-players-crossover-move-to-new-jersey.html.

The quote attributed to Dan Hurley is from www.vancouverbasketball.com in a May 7, 2015, article.

The stats about Thompson's offensive rebound count in the 2015–16 season are from the ESPN website (http://www.espn.com/nba/statistics/player/_/stat/rebounds/sort/avgOffensiveRebounds/year/2016).

The rise of the importance of the three-point shot has been written about in various places, but Celtics coach and president Red Auerbach's disdainful comments about what he considered a frivolous ratings grab are from Victor Mather, "How the N.B.A. 3 Point Shot Went from Gimmick to Game Changer," *New York Times* (January 20, 2016), http://www.nytimes.com/2016/01/21/sports/basketball/how-the-nba-3-point-shot-went-from-gimmick-to-game-changer.html. Anyone who has watched a compilation video of Steph Curry's three-pointers will disagree with Auerbach: they definitely make the game more exciting to watch.

Three-point shots also mean that smaller players, who aren't as handy inside the paint, can contribute more. See Charles Bethea, "Two of the World's Greatest Shooters Consider the Four-Point Shot," *New Yorker* (May 20, 2016) http://www.newyorker.com/news/sporting-scene/larry-bird-and-reggie-miller-consider-the-four-pointer.

The Stages of Change theory advanced by James Prochaska and his colleagues has become the dominant model today. From psychologists to business consultants, everyone uses it in their work. Prochaska has written several books, but the one that popularized his Stages of Change research is *Changing for Good: A Revolutionary Six-Stage Program for Overcoming Bad Habits and Moving Your Life Positively Forward* (New York: William Morrow, 1994).

The quote about Tristan Thompson making the switch to his right hand is from this interview and article by Michael Grange of SportsNet: http://www.sportsnet.ca/basketball/nba/michael-grange-tristan-thompson-making-history-by-switching-to-shoot-right-handed. Thompson's remarks about people sticking with what they know out of insecurity are from a *Sports Illustrated* interview in November 2012.

Chapter Ten: Finding the Meaning

Like most Canadians, I knew Maher Arar's story from watching it play out in the media. When I met him in person in 2016, I was struck by how gentle he was, and by his low-key sense of humour and flashing smile. He agreed to be interviewed for this book and then also for my television show. He was focused on his future and his new business venture, but I was fascinated by how at peace he seemed, despite his terrible ordeal. To tell his story, I relied on those interviews and on the book written by his wife, Monia Mazigh: *Hope and Despair: My Struggle to Free My Husband, Maher Arar* (Toronto: McClelland and Stewart, 2008). I also relied on several press reports, including the following: "The Story of Maher Arar: Unfolding US–Canada Police State," Global Research (December 29, 2005), http://www.globalresearch.ca/the-story-of-maher-arar-unfolding-us-canada-police-state/1664; Jonathon Gatehouse, "Maher Arar's Mind Cannot Forget," *Maclean's* (September 8, 2011), http://www.macleans.ca/news/canada/the-mind-cannot-forget; Tonda MacCharles, "RCMP Charges Syrian Colonel with Torture in Maher Arar Case," *Toronto Star*

(September 1, 2015), https://www.thestar.com/news/canada/2015/09/01/rcmp-to-charge-syrian-in-maher-arar-torture.html; Jeff Sallot, "How Canada Failed Citizen Maher Arar," *Globe and Mail* (September 19, 2006), http://www.theglobeandmail.com/news/national/how-canada-failed-citizen-maher-arar/article1103562; Sean Silcoff, "I'm a Tech Guy," *Globe and Mail* (April 23, 2016), http://www.theglobeandmail.com/report-on-business/careers/careers-leadership/human-rights-icon-maher-arar-is-back-to-doing-what-he-loves/article29714649; and Colby Itkowitz, "From Tortured Terrorist Suspect to Entrepreneur: How This Canadian Father Got His Life Back," *Washington Post* (April 27, 2016), https://www.washingtonpost.com/news/inspired-life/wp/2016/04/27/from-accused-terrorist-to-canadian-entrepreneur-maher-arar-is-finally-getting-his-life-back.

Barry Pokroy is a clinical psychologist who now specializes in organizational behaviour and the people inside businesses. Often that means helping them understand the stages of change and the best way to effect change. His company is called Circle and Square (http://www.farberfinancial.com/circle-and-square-canada).

INDEX